Electronic Test Instruments

HEWLETT-PACKARD PROFESSIONAL BOOKS

Electronic Test Instruments

—— Theory and Applications ——

Robert A. Witte

Hewlett-Packard Company

PTR Prentice Hall
Englewood Cliffs
New Jersey 07632

Library of Congress Cataloging-in-Publication Data

Witte, Robert A.
    Electronic test instruments : theory and applications / Robert A.
Witte.
      p. cm.
    Includes bibliographical references and index.
    ISBN 0-13-253147-X
    1. Electronic instruments. 2. Electronic measurements.
I. Title.
TK7878.4.W59 1993
621.3815′48—dc20   92-15942
                                                       CIP

Editorial production	**Copy editor:** *Sally Ann Bailey*
and interior design: *bookworks*	**Editor-in-Chief:** *Bernard Goodwin*
Acquisitions editor: *Karen Gettman*	**Prepress buyer:** *Mary McCartney*
Managing editor: *Sophie Papanikolaou*	**Manufacturing buyer:** *Susan Brunke*
Cover designer: *Lundgren Graphics, Ltd.*	

 Published by PTR Prentice Hall
A division of Simon & Schuster
Englewood Cliffs, New Jersey 07632

The publisher offers discounts on this book when ordered in bulk quantities.
For more information write:
    Special Sales/College Marketing
    Prentice Hall
    College Technical and Reference Division
    Englewood Cliffs, New Jersey 07632

Printed in the United States of America

10 9 8 7 6 5 4 3 2 1

ISBN 0-13-253147-X

Prentice-Hall International (UK) Limited, *London*
Prentice-Hall of Australia Pty. Limited, *Sydney*
Prentice-Hall Canada, Inc., *Toronto*
Prentice-Hall Hispanoamericana S.A., *Mexico*
Prentice-Hall of India Private Limited, *New Delhi*
Prentice-Hall of Japan, Inc., *Tokyo*
Simon & Schuster Asia Pte, Ltd., *Singapore*
Editora Prentice-Hall do Brasil, Ltda., *Rio De Janeiro*

To my family: Joyce, Sara, and Rachel

Contents

CHAPTER 2 VOLTMETERS, AMMETERS, AND OHMMETERS   **46**

CHAPTER 3 SIGNAL SOURCES   **69**

Preface

This book is for the student, technician, or engineer who understands basic electronics and wants to learn more about electronic measurements. To use instruments effectively, it is necessary to understand basic measurement theory and how it relates to practical measurements. Basic measurement theory includes such things as how a voltage's waveform relates to its frequency and how an instrument can affect the voltage that it is measuring. Ideally, it is desirable to *not* have to understand the internal operation of an instrument to use it. Although this ideal situation can be approached, it cannot be obtained completely. (One does not have to know how a gasoline engine works to drive an automobile. However, a driver should at least understand the function of the accelerator and brake pedals.)

Much of this material was originally published as *Electronic Test Instruments: A User's Sourcebook*. The original material has been updated and completely new sections have been added. The mathematical level was upgraded somewhat, but the more involved analyses appear in Appendix D. A complete chapter on logic analyzers, an important tool for digital design and test, is included.

To minimize dealing with the internal workings of an instrument, circuit models and conceptual block diagrams are used extensively. Circuit models take a "black box" approach to describing a circuit. In other words, the behavior of a complex circuit or instrument can be described adequately by conceptually replacing it with a much simpler circuit. This circuit model approach reduces the amount of detail that must be remembered and understood. Conceptual block diagrams show just enough of the inner workings of an instrument so that the reader can understand what the instrument is doing, without worrying about the details of how this is accomplished. In some instruments, most notably oscilloscopes, a greater understanding of the internal workings of the instrument is required since these factors impact how the instrument is used.

The traditional analog technologies are gradually yielding to digital techniques. However, a voltage measurement is still a voltage measurement whether an analog meter or a digital meter is used. Since the measurement is fundamentally the same, this book treats both technologies in a unified manner, while still highlighting the differences. Externally, both types of instruments do basically the same thing, with some important differences in how the reading is displayed. As appropriate, these differences are explained.

This book does not attempt to be (nor can it be) a substitute for a well-written instrument operating manual. The reader is not well served by a book that says "push this button, turn this knob" because the definition of the buttons and knobs will undoubtedly change with time. Instead, this book is a supplement or reference which provides the reader with a background in electronic instruments. Variations and improvements in instrument design cause each meter, oscilloscope, or function generator to be unique. However, they all have in common the fundamental measurement principles discussed in this book.

Chapter 1 addresses the basic measurement theory and fundamentals. Chapters 2 through 6 cover the mainstream instruments that the typical user will encounter (meters, signal sources, oscilloscopes, frequency counters, etc.). Some circuit concepts and how they relate to electronic measurements are discussed in Chapter 7. Chapter 8 introduces frequency domain instruments, with the emphasis on spectrum analyzers. Finally, Chapter 9 considers logic probes and logic analyzers.

The original motivation for this book was the author's experience teaching undergraduate courses in electrical engineering. Even students with a good background in electrical theory seem to have trouble relating the textbook concepts to what is observed in the laboratory. The concepts of the loading effect, grounding, and bandwidth are particularly troublesome. Hence, they are emphasized throughout the book. Although this book is not a textbook, it certainly could be used as supplemental reading for an engineering or technology laboratory course.

Acknowledgments

My thanks go to B&K-Precision/Dynascan Corporation, the Hewlett-Packard Company, and the Leader Instruments Corporation for providing photographs of their products.

My thanks also go to all the professors, engineers, and technicians with whom I have had the privilege of working, learning, and playing (not necessarily in that order). My sincere appreciation goes particularly to the people who assisted me in the preparation of this book: James Kahkoska, Jerry Murphy, Bill Spaulding, Scott Steven, Ken Wyatt, and Joyce Witte.

Robert A. Witte

Electronic Test Instruments

chapter one

—————— *Measurement Theory* ——————

To understand the operation and use of electronic instruments, it is necessary to review some of the electrical theory associated with electronic measurements. Although it is assumed that the reader understands basic electrical principles (voltage, current, Ohm's law, etc.), these principles will be reviewed here with special emphasis on how the theory relates to electronic measurement. With this approach, the theory is tackled up front to lay the groundwork for discussing the use and operation of electronic instruments. Most of the theoretical concepts apply to multiple types of measurements and instruments.

1.1 ELECTRICAL QUANTITIES

Before attempting to discuss electronic measurements, we must first make sure that the parameters to be measured are well understood. Appendix A contains a table of electrical parameters, their units of measure, and standard abbreviations. The standard electrical units can be modified by the use of prefixes (milli, kilo, etc.). Their use is also summarized in Appendix A. Our initial concern is with the measurement of voltage and current (with a slight emphasis on voltage); later we will include other parameters such as capacitance and inductance.

CURRENT (measured in units of amperes and often abbreviated to amps) is the flow of electrical CHARGE (measured in coulombs). The amount of charge is determined by the number of electrons moving past a given point. An electron has a negative charge of 1.602×10^{-19} coulombs, or equivalently, a coulomb of negative charge consists of 6.242×10^{18} electrons. The unit of current (the ampere) is defined as the number of coulombs of charge passing a given point in a second (one ampere equals one coulomb per second). The more charge that moves in a

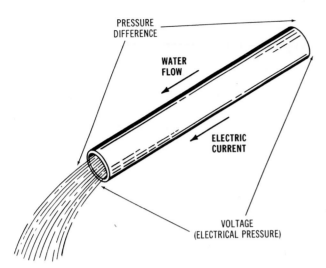

PRESSURE
DIFFERENCE

WATER
FLOW

ELECTRIC
CURRENT

VOLTAGE
(ELECTRICAL PRESSURE)

Figure 1-1   The water pipe analogy shows how water flow and pressure difference behave similarly to electrical current and voltage.

given time, the higher the current. Even though current is usually made up of moving electrons, the standard electrical engineering convention is to consider current to be the flow of positive charge.[1] With this definition, the current is considered to be flowing in the direction opposite of the electron flow (since electrons are negatively charged).

VOLTAGE (measured in volts), often referred to as electromotive force (EMF) or electrical potential, is the electrical force or pressure that causes the charge to move and the current to flow. Voltage is a relative concept; that is, voltage at a given point must be specified relative to some other point, which may be the system common or ground point.

An often-used analogy to electrical current is a water pipe with water flowing through it (Figure 1-1). The individual water molecules can be thought of as electrical charge. The amount of water flowing is similar to electrical current. The water pressure (presumably provided by some sort of external pump) corresponds to electrical pressure or voltage. In this case, the water pressure that interests us is actually the difference between the two pressures at each end of the pipe. If the pressure (voltage) is the same at both ends of the pipe, the water flow (current) is zero. On the other hand, if the pressure (voltage) is higher at one end of the pipe, water (current) flows away from the higher pressure end toward the lower pressure end.

Note that while the water flows THROUGH the pipe, the water pressure is ACROSS the pipe. In the same way, current flows THROUGH an electrical device but voltage (electrical pressure) exists ACROSS the device (Figure 1-2). This affects the way we connect measuring instruments depending on whether we are measuring voltage or current. For voltage measurement, the measuring instru-

---

[1]Current is also sometimes defined as electron flow (i.e., the current flows opposite to the direction used in this book).

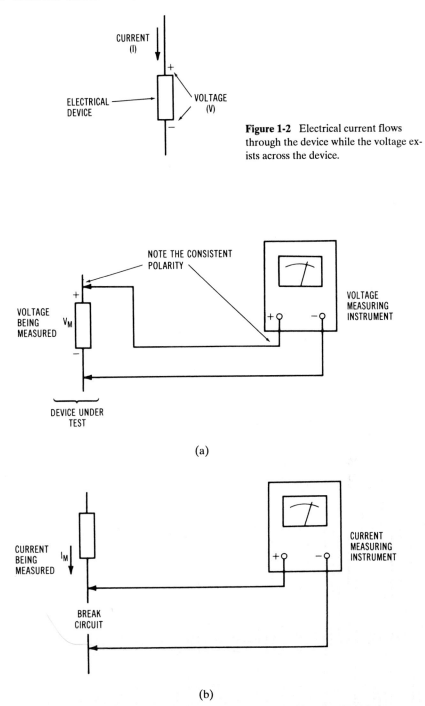

Figure 1-2 Electrical current flows through the device while the voltage exists across the device.

(a)

(b)

Figure 1-3 (a) Voltage measurements are made at two points. Note that the polarity of the measuring instrument is consistent with the polarity of the voltage being measured. (b) Current measurements are made by inserting the measuring instrument into the circuit so that the current flows through the instrument.

ment is connected in parallel with the two voltage points (Figure 1-3a). Two points must be specified when referring to a particular voltage, and two points are required for a voltage measurement (one of them may be the system ground). Measuring voltage at one point only is incorrect. Often we refer to a voltage at one point when the other point is implied to be the system common or ground point. This is an acceptable practice as long as the assumed second point is made clear. Current, similar to water flow, passes through a device or circuit. When measuring current, the instrument is inserted into the circuit that we are measuring (Figure 1-3b). (There are some exceptions to this, such as current probes discussed in Chapter 4.) The circuit is broken at the point the current is to be measured and the instrument is inserted. This results in the current being measured passing through the measuring instrument. To preserve accuracy for both voltage and current measurements, it is important that the measuring instrument not affect the circuit that is being measured.

1.2 RESISTANCE

A RESISTOR is an electrical device which obeys Ohm's law:

$$I = \frac{V}{R} \quad \text{or} \quad V = I \cdot R \quad \text{or} \quad R = \frac{V}{I}$$

Ohm's law simply states that the current through a resistor is proportional to the voltage across that resistor. Returning to the water pipe analogy, as the pressure (voltage) is increased, the amount of water flow (current) also increases. If the voltage is reduced, the current is reduced. The resistance, carrying the analogy further, is related inversely to the size of the water pipe. The larger the water pipe, the smaller the resistance to water flow. A large water pipe (small resistor) allows a large amount of water (current) to flow for a given pressure (voltage). The name "resistor" is due to the behavior of the device: it resists current. The larger the resistor, the more it resists and the smaller the current (again, assuming a constant voltage).

1.3 POLARITY

A small amount of care must be taken to ensure that voltages and currents calculated or measured have the proper direction associated with them. The standard engineering sign convention (definition of the direction of the current relative to the polarity of the voltage) is shown in Figure 1-4. A voltage source (V) is connected to a resistor (R) and some current (I) flows through the resistor. The direction of I is such that for a positive voltage source, a positive current will leave the voltage source at the positive terminal and enter the resistor at its most positive end. (Remember, current is taken to be the flow of positive charge. The electrons would actually be moving opposite

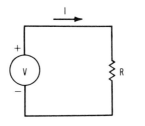

$$I = \frac{V}{R}$$

Figure 1-4 Ohm's law is used to compute the amount of current (I) that will result with voltage (V) and resistance (R).

the current.) Figure 1-3 shows the measuring instrument connected up in a manner consistent with the directions in Figure 1-4.

If the measuring instruments are connected up with the wrong polarity (backward) when making direct current measurements, the instrument will attempt to measure the proper value, but with the wrong sign (e.g., −5 volts instead of +5 volts). Typically, in digital instruments this is no problem. A minus sign is just added in front of the reading. In an analog instrument the reading will usually go off the scale, and if an electromechanical meter is used, it will deflect the meter in the reverse direction, often causing damage. It is best to consult the operating manual to understand the limitations of the particular instrument before attempting measurements.

Example 1-1

Calculate the amount of voltage across a 5-kΩ resistor if 3 mA of current is flowing through it.

Using Ohm's law,

$$V = I \cdot R = (0.003)(5000) = 15 \text{ volts}$$

1.4 DIRECT CURRENT

DIRECT CURRENT (DC) is the simplest form of current. Both current and voltage are constant with respect to time. A plot of DC voltage[2] versus time is shown in Figure 1-5. This plot should seem rather uninformative, since it shows the same voltage for all values of time, but it will contrast well with alternating current when it is introduced.

Both batteries and DC power supplies produce DC voltage. Batteries are available with quite a variety of voltage and current ratings. DC power supplies convert AC voltages into DC voltages and are discussed in Chapter 7.

---

[2]Even though the term DC specifically states direct current, it is used to describe both voltage and current. This leads to commonly used, but self-contradictory terminology such as "DC voltage," which means, literally, "direct current voltage."

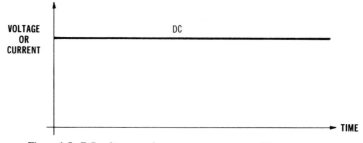

Figure 1-5 DC voltages and currents are constant with respect to time.

1.5 POWER

POWER is the rate at which energy flows from one circuit to another. For DC voltages and currents, power is simply the voltage times the current and the unit is the watt:

$$P = V \cdot I$$

Using Ohm's law and some simple math, the relationships in Table 1-1 can be developed. Notice that power depends on both current and voltage. There can be high voltage but no power if there is no path for the current to flow. Or there could be a large current flowing through a device with zero volts across it, also resulting in no power being received by the device.

TABLE 1-1 BASIC EQUATIONS FOR DC VOLTAGE, DC CURRENT, RESISTANCE, AND POWER.

$V = I \cdot R$	Ohm's law
$I = V/R$	Ohm's law
$R = V/I$	Ohm's law
$P = V \cdot I$	Power equation
$P = V^2/R$	Power in resistor
$P = I^2 \cdot R$	Power in resistor

1.6 ALTERNATING CURRENT

ALTERNATING CURRENT (AC), as the name implies, does not remain constant with time as direct current does, but instead changes direction (alternates) at some frequency. The most common form of AC is the sine wave, as shown in Figure 1-6. The current or voltage starts out at zero, becomes positive for one-half of the cycle and then passes through zero to become negative for the second half of the cycle. This cycle then repeats continuously. The sine wave can be described mathematically as a function of time.

$$v(t) = V_{0-P} \sin(2 \pi f t)^3$$

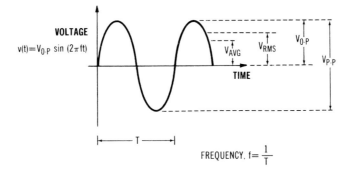

Figure 1-6 The most common form of alternating current is the sine wave. The voltage of the waveform can be described by the RMS value, zero-to-peak value, or the peak-to-peak value.

The length of the cycle (in seconds) is called the PERIOD and is represented by the symbol T. The FREQUENCY, f, is the reciprocal of the period and is measured in units of hertz, which is equivalent to cycles per second.

$$f = \frac{1}{T}$$

The frequency indicates how many cycles the sine wave completes in one second. For example, the standard AC power line voltage in the United States has a frequency of 60 Hz, which means that the voltage goes through 60 complete cycles in 1 second. The period of a 60-Hz sine wave is $T = 1/f = 1/60 = 0.0167$ sec.

Sometimes the sine wave equation is presented in the following form

$$v(t) = V_{0-P} \sin(\omega t)$$

where ω is the radian frequency, with units of radians per second.

By comparing the two equations for the sinusoidal voltage, we can see that

$$\omega = 2\pi f$$

and

$$f = \frac{\omega}{2\pi}$$

Since the sine wave is not constant with time, it is not immediately obvious how to describe its voltage. Sometimes the voltage is positive, sometimes it is negative, and twice every cycle it is zero. This problem does not exist with DC, since it is always a constant value. Figure 1-6 shows four different ways of referring to AC voltage. The zero-to-peak value (V_{0-P}) is simply the maximum voltage that the sine wave reaches. Similarly, the peak-to-peak value (V_{P-P}) is measured from the maximum positive voltage to the most negative voltage. For a sine wave, V_{P-P} is always twice V_{0-P}.

[3]$\pi = 3.14159$ (approximately).

1.7 RMS VALUE

Another way of referring to AC voltage is the RMS value (V_{RMS}). RMS is an abbreviation for root mean square, which indicates the mathematics behind calculating the value of arbitrary waveforms. To calculate the RMS value of a waveform, the waveform is first squared at every point. Then, the average or mean value of this squared waveform is found. Finally, the square root of the mean value is taken to produce the RMS (root of the mean of the square) value.

Mathematically, the RMS value of a waveform can be expressed as

$$V_{RMS} = \sqrt{\frac{1}{T}\int_{t_0}^{t_0 + T} v^2(t)\, dt}$$

Determining the RMS value from the zero-to-peak value (or vice versa) can be difficult due to the complexity of the RMS operation. Fortunately, the relationship has been computed for the sine wave and is very straightforward. See Appendix D for the detailed analysis.

$$V_{RMS} = \frac{1}{\sqrt{2}} V_{0-P} = 0.707\, V_{0-P}$$

This relationship is valid only for a sine wave and does *not* hold for other waveforms.

1.8 AVERAGE VALUE

Finally, voltage is sometimes defined using an average value. Strictly speaking, the average value of a sine wave is zero because the waveform is positive for one-half cycle and is negative for the other half. Since the two halves are symmetrical, they cancel out when they are averaged together. So, on the average, the waveform voltage is zero.

Another interpretation of average value is to assume that the waveform has been full-wave rectified. Mathematically this means that the absolute value of the waveform is used (i.e., the negative portion of the cycle has been treated as being positive).

$$V_{AVG} = \frac{1}{T}\int_{t_0}^{t_0 + T} |v(t)|\, dt$$

This is exactly what some measurement instruments do to handle AC waveforms, so that is what will be considered here. Unless otherwise indicated, V_{AVG} will mean the full-wave rectified average. These averaging steps have been shown in Figure 1-7. Figure 1-7a shows a sine wave that is to be full-wave rectified. Figure 1-7b shows the resulting full-wave rectified sine wave. Whenever the original waveform becomes negative, full-wave rectification changes the sign and converts the voltage into a positive waveform with the same amplitude. Graphically, this can be described as folding the negative half of the waveform up onto the positive half, resulting in a

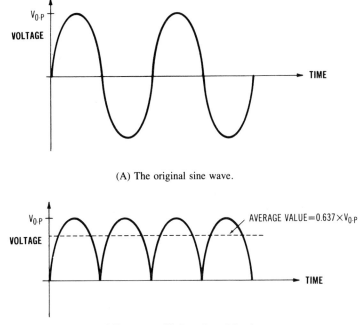

(A) The original sine wave.

(B) The full-wave rectified version of the sine wave.

Figure 1-7 The operations involved in finding the full-wave average value of a sine wave. (a) The original sine wave. (b) The full-wave rectified version of the sine wave. The negative half cycle is folded up to become positive. The resulting waveform is then averaged.

humped sort of waveform. Now the average value can be determined and is plotted in Figure 1-7b. The relationship between V_{AVG} and V_{0-P} depends on the shape of the waveform. For a sine wave

$$V_{\text{AVG}} = \frac{2}{\pi} V_{0-P} = 0.637\, V_{0-P} \qquad \text{(sine wave)}$$

The details of the analysis are shown in Appendix D.

1.9 CREST FACTOR

The ratio of the zero-to-peak value to the RMS value of the waveform is known as the CREST FACTOR. The crest factor is a measure of how high the waveform peaks, relative to its RMS value. The crest factor of a waveform is important in some measuring instruments. Waveforms with very high crest factors require the measuring instrument to tolerate very large peak voltages while simultaneously measuring the much smaller RMS value.

$$\text{crest factor} = \frac{V_{0-P}}{V_{\text{RMS}}}$$

Example 1-2

What is the crest factor of a sine wave?

For a sine wave, $V_{RMS} = 0.707 \ V_{0-P}$, so the crest factor is $1/0.707 = 1.414$. The sine wave has a relatively low crest factor. Its zero-to-peak value is not that much greater than its RMS value.

The PEAK-TO-AVERAGE RATIO, also known as the AVERAGE CREST FACTOR, is similar to the crest factor except that the average value of the waveform is used in the denominator of the ratio. It also is a measure of how high the peaks of the waveform are compared to its average value.

Although the preceding discussion has referred to AC voltages, the same concepts apply to AC currents. That is, AC currents can be described by their zero-to-peak, peak-to-peak, RMS, and average values.

1.10 PHASE

The voltage specifies the amplitude or height of the sine wave, and the frequency (or the period) specifies how often the sine wave completes a cycle. But two sine waves of the same frequency may not cross zero at the same time. Therefore, the PHASE of the sine wave is used to define its position on the time axis. The unit of phase is the degree, with one cycle of a sine wave divided up into 360 degrees of phase.

The mathematical definition of the sine wave can be modified to include a phase term.

$$v(t) = V_{0-P} \sin (2 \ \pi ft + \theta)$$

As written, the equation implies that phase is absolute. That is, there is some instant in time when $t = 0$ to which the phase angle is referenced. In practice there usually is no such universal time, and phase is a relative concept. In other words, we can normally talk about phase between two sine waves, but not the phase of a single, isolated sine wave (unless some other sort of time reference is supplied).

For example, the two sine waves in Figure 1-8a are separated by one-fourth of a cycle (they reach their maximum values one-fourth of a cycle apart). Since one cycle equals 360 degrees, the two sine waves have a phase difference of 90 degrees. To be more precise, the second sine wave is 90 degrees behind the first, or, equivalently, the second sine wave has a phase of -90 degrees relative to the first sine wave. (The first sine wave has a $+90$-degree phase relative to the second sine wave.) It is also correct to say that the first sine wave leads the second one by 90 degrees or that the second sine wave lags the first one by 90 degrees. All these statements specify the same phase relationship.

Figure 1-8b shows two sine waves that are one-half cycle (180 degrees) apart. This is the special case where one sine waves is the negative of the other. The phase relationship between two sine waves simply defines how far one sine wave is shifted with respect to the other. When a sine wave is shifted by 360 degrees, it is shifted a

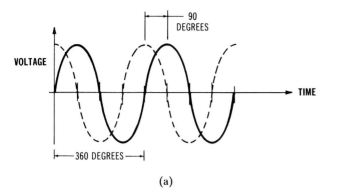

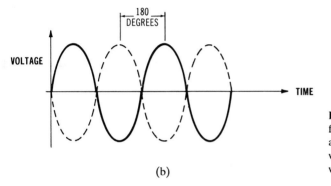

Figure 1-8   The phase of a sine wave defines its relative position on the time axis. (a) The phase between the two sine waves is 90 degrees. (b) The two sine waves are 180 degrees out of phase.

complete cycle and is indistinguishable from the original waveform. Because of this, phase is usually specified over a 360-degree range, typically −180 degrees to +180 degrees.

1.11 AC POWER

The average power dissipated by a resistor with an AC voltage across it is given by

$$P = V_{\text{RMS}} I_{\text{RMS}} = \frac{V^2_{\text{RMS}}}{R} = I^2_{\text{RMS}} \cdot R$$

This relationship holds for any waveform as long as the RMS value of the voltage and current are used. Note that these equations have the same form as the DC case, which is one of the reasons for using RMS values. The RMS value is often called the EFFECTIVE VALUE, since an AC voltage with a given RMS value has the same effect (in terms of power) that a DC voltage with that same value. (A 10-volt RMS AC voltage and a 10-volt DC voltage both supply the same power, 20 watts, to a 5-ohm resistor.) In addition, two AC waveforms that have the same RMS value will cause the same power to be delivered to a resistor. This is *not* true for other voltage descriptions such as zero-to-peak and peak-to-peak. Thus, RMS is the great equalizer (with respect to power).

Example 1-3.

The standard line voltage in the United States is approximately 120 volts RMS. What are the zero-to-peak, peak-to-peak, and full-wave rectified average voltages? How much power is supplied to a 200-ohm resistor connected across the line?

$$V_{RMS} = 0.707 \, V_{0-P}$$

so

$$V = \frac{V_{RMS}}{0.707} = \frac{120}{0.707} = 169.7 \text{ volts}$$

$$V = 2V_{0-P} = 2(169.7) = 339.4 \text{ volts}$$

$$V_{AVG} = 0.637 V_{0-P} = 0.637 \times 169.7 = 108.1 \text{ volts}$$

$$P = \frac{V_{RMS}^2}{R} = \frac{120^2}{200} = 72 \text{ watts}$$

1.12 NONSINUSOIDAL WAVEFORMS

There are other AC voltage and current waveforms besides sine waves that are commonly used in electronic systems. Since most of the concepts relating to waveforms apply to both voltage and current waveforms, voltage terminology will be used with the understanding that the same concepts are valid for current waveforms. Voltages are usually, but not always, more easily measured than currents, mainly due to the fact that voltage measurements can be made in parallel with the device being tested without disrupting the circuit paths.

Some of the more common waveforms are shown in Table 1-2. Note that the values of V_{RMS} and V_{AVG} (relative to V_{0-P}) are unique for each waveform. The first three waveforms are symmetrical about the horizontal axis, but the half-sine wave and the pulse train are always positive. All these waveforms are PERIODIC because they repeat the same cycle or period continuously.

An example will help to emphasize the utility of RMS voltages when dealing with power in different waveforms.

Example 1-4

A sine wave voltage and a triangle wave voltage are each connected across two separate 300-ohm resistors. If both waveforms deliver 2 watts (average power) to their respective resistors, what are the RMS and zero-to-peak voltages of each waveform?

The two waveforms deliver the same power to identical resistors, so their RMS voltages must be the same. (This is *not* true of their zero-to-peak values.)

$$P = \frac{V_{RMS}^2}{R}$$

$$V_{RMS} = \sqrt{P \cdot R} = \sqrt{2 \cdot 300} = 24.5 \text{ volts RMS}$$

From Table 1-2, for the sine wave

TABLE 1-2 TABLE OF WAVEFORMS WITH PEAK-TO-PEAK VOLTAGE (V_{P-P}), RMS VOLTAGE (V_{RMS}), ZERO-TO-PEAK VOLTAGE (V_{0-P}), AND CREST FACTOR FOR EACH WAVEFORM. V_{AVG} IS THE FULL-WAVE RECTIFIED AVERAGE VALUE OF THE WAVEFORM.

Waveform	V_{P-P}	V_{RMS}	V_{AVG}	Crest Factor
SINE WAVE	$2\,V_{0-P}$	$\frac{1}{\sqrt{2}}V_{0-P}$ or $0.707\,V_{0-P}$	$\frac{2}{\pi}V_{0-P}$ or $0.637\,V_{0-P}$	$\sqrt{2}$ or 1.414
SQUARE WAVE	$2\,V_{0-P}$	V_{0-P}	V_{0-P}	1
TRIANGLE WAVE	$2\,V_{0-P}$	$\frac{1}{\sqrt{3}}V_{0-P}$ or $0.577\,V_{0-P}$	$\frac{1}{2}\,V_{0-P}$	$\sqrt{3}$ or 1.732
HALF SINE WAVE	V_{0-P}	$\frac{1}{2}\,V_{0-P}$	$\frac{1}{\pi}\,V_{0-P}$ or $0.318\,V_{0-P}$	2
PULSE TRAIN	V_{0-P}	$\sqrt{\frac{\tau}{T}}\,V_{0-P}$	$\frac{\tau}{T}\,V_{0-P}$	$\sqrt{\frac{T}{t}}$

$$V_{RMS} = 0.707\,V_{0-P}$$

$$V_{0-P} = \frac{V_{RMS}}{0.707} = \frac{24.5}{0.707} = 34.7 \text{ volts}$$

For the triangle wave

$$V_{RMS} = 0.577\,V_{0-P}$$

$$V_{0-P} = \frac{V_{RMS}}{0.577} = \frac{24.5}{0.577} = 42.5 \text{ volts}$$

So for the triangle wave to supply the same average power to a resistor, it must reach a higher peak voltage than the sine wave.

1.13 HARMONICS

Periodic waveforms, except for absolutely pure sine waves, contain frequencies called HARMONICS. Harmonic frequencies are integer multiples of the original or FUNDAMENTAL frequency.

$$f_n = n \cdot f_{\text{fundamental}}$$

For example, a nonsinusoidal waveform with a fundamental frequency of 1 kHz would produce harmonics at 2 kHz, 3 kHz, 4 kHz, and so on. There can be any number of harmonics, including infinity, but usually there is a practical limitation on how many need be considered. Each harmonic may have its own unique phase relative to the fundamental.

Harmonics are present because periodic waveforms, regardless of shape, can be broken down mathematically into a series of sine waves. But this behavior is more than just mathematics. It is as if the physical world regards the sine wave as the purest, most simple sort of waveform, with all other periodic waveforms being made up of collections of sine waves. The result is that a periodic waveform (such as a square wave) is exactly equivalent to a series of sine waves.

1.14 SQUARE WAVE

Consider the square wave shown in Figure 1-9a. Square waves are made up of the fundamental frequency plus an infinite number of odd harmonics.[4] In theory, it takes every one of those infinite number of harmonics to create a true square wave. In practice, the fundamental and several harmonics will approximate a square wave. In Figure 1-9b, the fundamental, 3rd harmonic, and 5th harmonic are plotted. (Remember, the even harmonics of a square wave are zero.) Note that the amplitude of each higher harmonic is less than the previous one, so the highest harmonics may be small enough to be ignored. When these harmonics are added up, they produce a waveform that resembles a square wave. Figure 1-9c shows the waveform that results from combining just the fundamental and the 3rd harmonic. Already the waveform starts to look somewhat like a square wave. (Well, a least a little bit.) The fundamental plus the 3rd and 5th harmonics is shown in Figure 1-9d. It is a little more like a square wave. Figures 1-9e and 1-9f each add another odd harmonic to our square wave approximation and each one resembles a square wave more closely than the previous waveform. If all the infinite number of harmonics were included, the resulting waveform would be a perfect square wave. So the quality of the square wave is limited by the number of harmonics present.

The amplitude of each harmonic must be just the right value for the resulting wave to be a square wave. In addition, the phase relationships between the harmonics must also be correct. If the harmonics are delayed in time by unequal amounts, the square wave will take on a distorted look even though the amplitudes of the harmonics may be correct. This phenomenon is used to advantage in square wave testing of amplifiers as discussed in Chapter 5. It is theoretically possible to construct a square wave electronically by connecting a large number of sine wave generators together such that each one contributes the fundamental frequency or a harmonic with just the right amplitude. In practice, this may be difficult because the frequency and phase of each oscillator must also be precisely controlled.

So far, waveforms have been characterized using a voltage (or current) versus time plot, which is known as the TIME DOMAIN representation. Another way of

---

[4]For a mathematical derivation of the frequency content of a square wave, see Witte (1991).

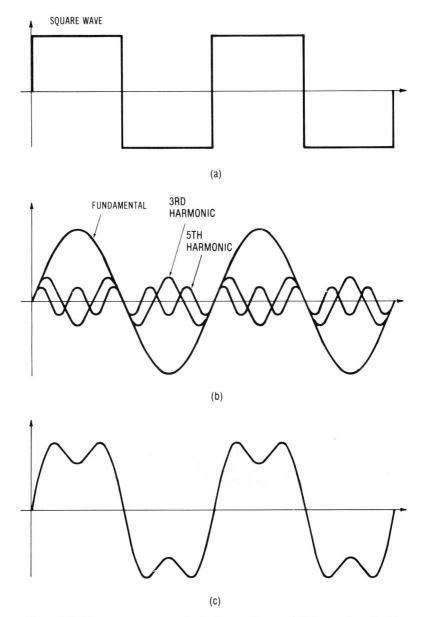

Figure 1-9 The square wave can be broken up into an infinite number of odd harmonics. The more harmonics that are included, the more the waveform approximates a square wave. (a) The original square wave. (b) The fundamental, 3rd harmonic, and 5th harmonic. (c) The fundamental plus 3rd harmonic. (c) The fundamental plus the 3rd and 5th harmonics. (d) The fundamental plus the 3rd, 5th, and 7th harmonics. (e) The fundamental plus the 3rd, 5th, 7th, and 9th harmonics.

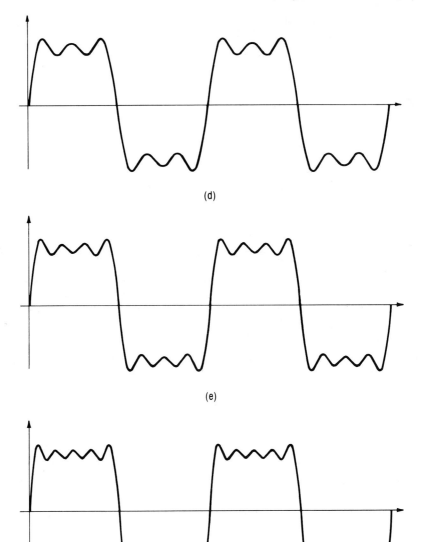

(d)

(e)

(f)

Figure 1-9   *(Continued)*

describing the same waveform is with frequency on the horizontal axis and voltage on the vertical axis. This is known as the FREQUENCY DOMAIN representation or SPECTRUM of the waveform. In the frequency domain representation, a vertical line (called a SPECTRAL LINE) indicates a particular frequency that is present (the fundamental or a harmonic). The height of each spectral line corresponds to the amplitude of that particular harmonic. A pure sine wave would be represented by one

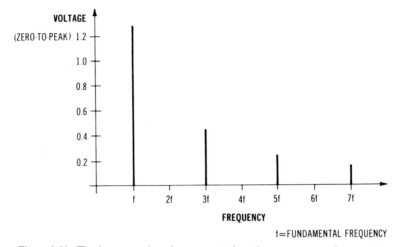

Figure 1-10 The frequency domain representation of a square wave, shown out to the 7th harmonic.

single spectral line. Figure 1-10 shows the frequency domain representation of a square wave. Notice that only the fundamental and odd harmonics are present and that each harmonic is smaller than the previous one. Understanding the spectral content of waveforms is important because the measurement instrument must be capable of operating at the frequencies of the harmonics (at least the ones that are to be included in the measurement).

1.15 PULSE TRAIN

The pulse train, or repetitive pulse, is a common signal in digital systems (Figure 1-11). It is similar to the square wave, but it does not have both positive and negative values. Instead it has two possible values: V_{0-P} and 0 volts. The square wave spends 50 percent of the time at its positive voltage and 50 percent of the time at its negative voltage. This is referred to as a 50 percent duty cycle. The pulse train's duty cycle may be any value between 0 and 100 percent and is defined by the following equation:

$$\text{duty cycle} = \tau/T$$

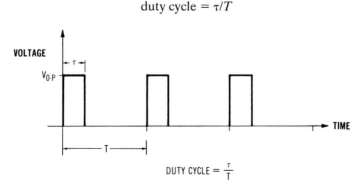

Figure 1-11 The pulse train is a common waveform in digital systems. The duty cycle describes the percentage of time that the waveform is at the higher voltage.

where τ is the length of time that the waveform is high and T is the period of the waveform.

The pulse train generates harmonic frequencies with amplitudes that are dependent on the duty cycle. A frequency domain plot of a typical pulse train (duty cycle = 25%) is shown in Figure 1-12. The envelope of the harmonics has a distinct humped shape which equals zero at integer multiples of $1/\tau$. Most of the waveform's energy is contained in the harmonics falling below this $1/\tau$ point. Therefore, it is often used as a rule of thumb for the bandwidth of the waveform.

Table 1-3 summarizes the harmonic content for each of the various waveforms with $V_{0-P} = 1$ volt. In all the waveforms (except the sine wave), there are an infinite number of harmonics, but the amplitudes of the harmonics tend to decrease as the harmonic number increases. At some point, the higher harmonics can be ignored for practical systems because they are so small. As a measure of how wide each waveform is in the frequency domain, the last column lists the number of significant harmonics (ones that are at least 10 percent of the fundamental). Notice that the amplitude value of the fundamental frequency component changes depending on the waveform, even though all the waveforms have $V_{0-P} = 1$ volt. The larger the number of significant harmonics, the wider the signal is in the frequency domain. The sine wave, of course, only has one significant harmonic (the fundamental can be considered the "first" harmonic).

There are two important concepts to be obtained from the previous discussion of harmonics.

1. The more quickly a waveform transitions between its minimum and maximum values, the more significant harmonics it will have and the higher the frequency content of the waveform. For example, the square wave (which has abrupt voltage changes) has more significant harmonics than does the triangle wave (which does not change nearly as quickly).
2. The narrower the width of a pulse, the more significant harmonics it will have and the higher the frequency content of the waveform.

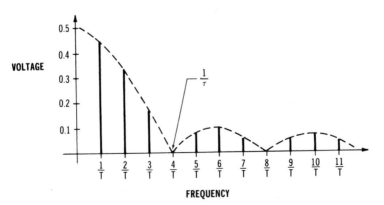

Figure 1-12   The frequency domain plot for a pulse train with a 25 percent duty cycle. The envelope of the harmonics falls to zero when the frequency equals $1/\tau$.

TABLE 1-3 TABLE OF HARMONICS FOR A VARIETY OF WAVEFORMS. ALL WAVEFORMS HAVE A ZERO-TO-PEAK VALUE OF 1. THE NUMBER OF SIGNIFICANT HARMONICS COLUMN LISTS THE HIGHEST HARMONIC WHOSE AMPLITUDE IS AT LEAST 10 PERCENT OF THE FUNDAMENTAL.

Waveform	Fund	Harmonics 2nd	3rd	4th	5th	6th	7th	Sig Harm (10%)	Equation
Sine wave	1.000	0.000	0.000	0.000	0.000	0.000	0.000	1	
Square	1.273	0.000	0.424	0.000	0.255	0.000	0.182	9	$\dfrac{4}{n\pi}$ for odd n
Triangle	0.811	0.000	0.090	0.000	0.032	0.000	0.017	3	$\dfrac{8}{n^2\pi^2}$ for odd n
Pulse (50% duty cycle)	0.637	0.000	0.212	0.000	0.127	0.000	0.091	9	$\dfrac{2}{n\pi}\sin\left(\dfrac{n\pi}{2}\right)$
Pulse (25% duty cycle)	0.450	0.318	0.150	0.000	0.090	0.105	0.064	14	$\dfrac{2}{n\pi}\sin\left(\dfrac{n\pi}{4}\right)$
Pulse (10% duty cycle)	0.197	0.187	0.172	0.151	0.127	0.101	0.074	26	$\dfrac{2}{n\pi}\sin\left(\dfrac{n\pi}{10}\right)$

Both statements should make intuitive sense if one considers that high-frequency signals change voltage faster than do lower-frequency signals. The two conditions just cited both involve waveforms changing voltage in a more rapid manner. Therefore, it makes sense that waveforms that must change rapidly (those with short rise times or narrow pulses) will have more high frequencies present in the form of harmonics.

Example 1-5

What is the highest frequency that must be included in the measurement of a triangle wave that repeats every 50 µsec (assuming that harmonics less than 10 percent of the fundamental can be ignored)?

The fundamental frequency $f = 1/T = 1/50$ µsec $= 20$ kHz. From Table 1-3, the number of significant harmonics for a triangle wave (using the 10 percent criterion) is 3. The highest frequency that must be included is $3f = 3(20$ kHz$) = 60$ kHz.

1.16 COMBINED COMBINED DC AND AC

In many cases, the voltage waveform present can be best thought of as being a combination of direct current and alternating current. For example, in transistor circuits, an AC waveform is often superimposed on a DC bias voltage as shown in Figure 1-13. The DC component (a straight horizontal line) and the AC component (a sine wave) combine to form the new waveform.

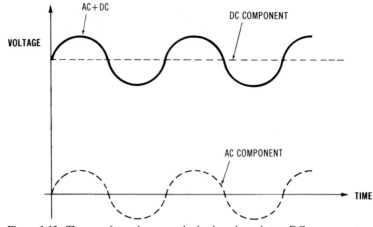

Figure 1-13 The waveform shown can be broken down into a DC component and an AC component. The DC component is just the average value of the original waveform.

The DC value of a waveform is also just the waveform's average value. In Figure 1-13, the waveform is above its DC value half of the time and below the DC value the other half, so on the average, the voltage is just the DC value.

Examine the 50 percent duty cycle pulse train shown in Figure 1-14. Notice that the waveform is always positive (it never goes below zero volts). The average value is, therefore, greater than zero. The waveform spends half of the time at V_{0-P} and the

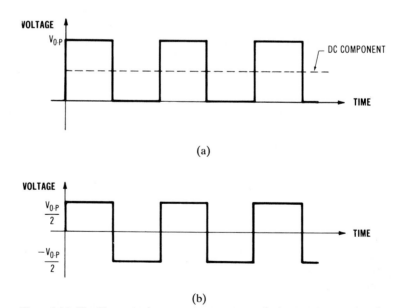

Figure 1-14 The 50 percent duty cycle pulse train is equivalent to a square wave plus a DC voltage. (a) The pulse train with DC component shown. (b) The square wave that results when the DC component is removed from the pulse train.

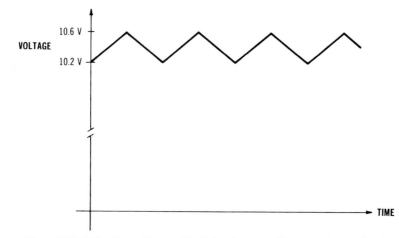

Figure 1-15 DC voltage with a residual triangle wave voltage superimposed on top of it (Example 1-6).

other half at 0, so the average (DC) value $= (V_{0-P} + 0)/2 = \frac{1}{2} V_{0-P}$. The AC component left over when the DC component is removed is a square wave with half the zero-to-peak value of the original pulse train. In summary then, the 50 percent duty cycle pulse train is equivalent to a square wave of half the zero-to-peak voltage plus a $\frac{1}{2}V_{0-P}$ DC component.

Example 1-6

A DC power supply has some residual AC riding on top of its DC component, as shown in Figure 1-15. Assuming that the AC component can be measured independently of the DC component, determine both the DC and AC values that would be measured (give the RMS value for the AC component).

The DC value is simply the average value of the waveform. Since the AC component is symmetrical, the average value can be calculated:

$$\text{DC value} = \frac{10.6 + 10.2}{2} = 10.4 \text{ volts}$$

If the DC is removed from the waveform, a triangle wave with $V_{0-P} = 0.2$ volts is left. For a triangle wave

$$V_{\text{RMS}} = 0.577 \, V_{0-P} = 0.115 \text{ volts RMS}$$

1.17 MODULATED SIGNALS

Sometimes sine waves are modulated by another waveform. For example, communication systems use this technique to superimpose low-frequency (voice or data) signals onto a high-frequency carrier which can be transmitted long distances. This modulation is performed by modifying some parameter of the original sine wave (called the CARRIER), depending on the value of the modulating waveform. In this

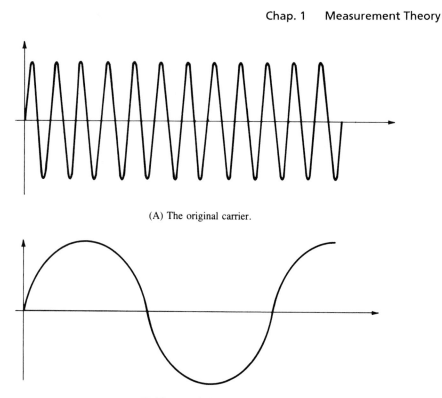

(A) The original carrier.

(B) The modulating signal.

Figure 1-16 Amplitude and frequency modulation. (a) The original carrier. (b) The modulating signal. (c) The resulting signal if amplitude modulation is used. (d) The resulting signal if frequency modulation is used.

way, information from the modulating waveform is transferred to the carrier. Any characteristic of the carrier can be used. The most common forms of modulation are amplitude, frequency, and phase modulation.

In amplitude modulation (AM), the height or amplitude of the carrier is determined by the modulating waveform. Figure 1-16a shows the original sine wave carrier with constant amplitude, Figure 1-16b is the modulating waveform, and Figure 1-16c is the modulated carrier with the modulating waveform determining the amplitude of the carrier. The modulating waveform can be seen as the envelope of the modulated waveform. When the modulating waveform increases, the amplitude of the carrier increases. When the modulating waveform decreases, the amplitude of the carrier decreases.

Frequency modulation (FM) also modulates a sine wave carrier, but the amplitude of the carrier remains constant while the frequency of the carrier changes. When the modulating waveform increases, the carrier frequency (the number of cycles per second) increases. When the modulating waveform decreases, the instantaneous carrier frequency decreases. A frequency modulated signal is shown in Figure 1-16d.

Phase modulation is very similar to frequency modulation. The carrier amplitude remains constant, and the phase is changed according to the modulating wave-

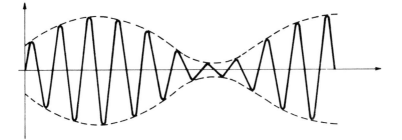

(C) The resulting signal if amplitude modulation is used.

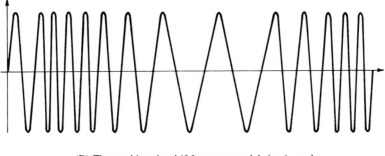

(D) The resulting signal if frequency modulation is used.

Figure 1-16 *(Continued)*

form. Since phase and frequency are closely related for our purposes, the effect of phase modulation is very similar to frequency modulation. In fact, phase and frequency modulation may be indistinguishable in a practical measurement system. They are generally considered to be variants of a type of modulation called ANGLE MODULATION.

Both AM and FM are used in a wide variety of communication systems, including standard radio broadcast stations. The radio station superimposes the modulating waveform (the data, voice, or music to be broadcast) onto the radio frequency carrier using either AM or FM. At the receiving end, the listener's radio receiver extracts the modulation from the incoming signal to recover the desired data, voice, or music information.

Modulating a carrier, either with AM or FM, has an effect in the frequency domain (Figure 1-17). With no modulation, the carrier is just a pure sine wave and exists only at one frequency (Figure 1-17a). When the modulation is added, the carrier is accompanied by a band of other frequencies, called SIDEBANDS. The exact behavior of these sidebands depends on the level and type of modulation, but in general the sidebands spread out on one or both sides of the carrier (Figure 1-17b). Often they are relatively close to the carrier frequency, but in some cases such as wideband FM, the sidebands spread out much farther. In all cases, modulation has the effect of spreading the carrier out in the frequency domain, occupying a wider frequency space.

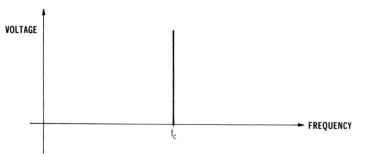

(A) The unmodulated carrier is a single spectral line.

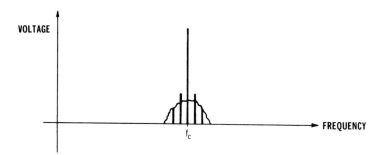

(B) Modulation on the carrier spreads the signal out in the frequency domain.

Figure 1-17   The effect of modulation in the frequency domain is to spread out the signal by creating sidebands. (a) The unmodulated carrier is a single spectral line. (b) Modulation on the carrier spreads the signal out in the frequency domain. (c) An amplitude modulated carrier (with single sine wave modulation) has a single pair of sidebands. (d) A frequency modulated carrier (with single sine wave modulation) may have many pairs of sidebands.

When a carrier is amplitude modulated by a single sine wave with frequency f_M, two modulation sidebands appear (offset by f_M) on both sides of the carrier (Figure 1-17c). Carriers that are frequency modulated may have a considerably larger number of sidebands. With a modulating frequency of f_M, the sidebands appear at integer multiples of f_M away from the carrier frequency. In the general case, FM is a wideband form of modulation, in that the modulated carrier has sidebands that extend out farther than f_M away from the carrier. If the modulation occurs at a low level, FM may be narrowband, with sidebands very similar to the AM case. For both modulation schemes, when the modulating signal is complex (voice waveforms, multiple sine waves, etc.), the sidebands are also more complex.

Radio systems represent intentional use of modulation. Sometimes modulation is produced due to imperfections in practical circuits. In this case, the modulation is either undesirable or at least unnecessary. The effect in the frequency domain is still the same, however, so a sine wave with some residual modulation on it has sidebands in the frequency domain, causing the signal to spread out.

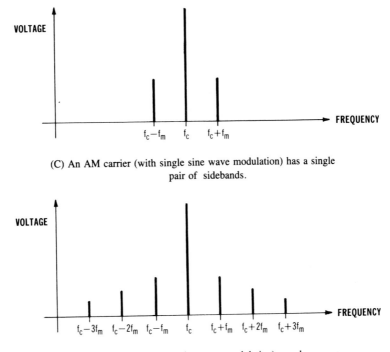

(C) An AM carrier (with single sine wave modulation) has a single pair of sidebands.

(D) An FM carrier (with single sine wave modulation) may have many pairs of sidebands.

Figure 1-17 *(Continued)*

1.18 DECIBELS

The decibel (dB) is sometimes used to express electrical quantities in a convenient form. The definition of the decibel is based on the ratio of two power levels (log indicates the base 10 logarithm):

$$dB = 10 \log\left(\frac{P_2}{P_1}\right)$$

Since $P = V^2/R$ (assuming RMS voltage),

$$dB = 10 \log\left(\frac{V_2^2 / R_2}{V_1^2 / R_1}\right)$$

And if $R_1 = R_2$ (the resistances involved are the same),

$$dB = 10 \log\left(\frac{V_2}{V_1}\right)^2$$

$$= 20 \log\left(\frac{V_2}{V_1}\right)$$

The voltages, V_1 and V_2, are normally RMS voltages. If the two waveforms are the same shape (for instance, if they are both sine waves), then the voltages can be expressed as zero-to-peak or peak-to-peak. Since many of our measurements are voltage measurements, calculating dB using voltages is the most convenient method. It should be pointed out that strictly speaking, the voltage equation is valid only if the two resistances involved are the same. For instances if the input voltage and output voltage of an amplifier are being compared and the input resistance and output resistance are different, they must be accounted for when calculating decibels. In practice, however, this is often ignored. For instance, in operational amplifier circuits, the input impedance is very large while the output impedance is quite small. Usually, the important parameter in this type of circuit is voltage gain (and not power gain). In this type of system, the decibel equation for voltage is used almost exclusively.

To convert dB values back into a voltage or power ratio, a little reverse math is required:

$$\frac{P_2}{P_1} = 10^{\mathrm{dB}/10}$$

$$\frac{V_2}{V_1} = 10^{\mathrm{dB}/20}$$

The usefulness of decibels is sometimes questioned, but decibels are widely used and must be understood by instrument users for that reason alone. In addition, they have at least two characteristics which make them very convenient:

1. Decibels compress widely varying electrical values onto a more manageable logarithmic scale. The range of powers extending from 100 watts down to 1 microwatt is a ratio of 100,000,000 to 1 but is expressed in dB as only 80 dB.
2. Gains and losses through circuits such as attenuators, amplifiers, filters, when expressed in dB can be added together to produce the total gain or loss. To perform the equivalent operation without dBs requires multiplication.

Some cardinal decibel values are worth pointing out specifically:

- 0 dB corresponds to a ratio of 1 (for both voltage and power). A circuit that has 0 dB gain or 0 dB loss has an output equal to the input.
- 3 dB corresponds to a power ratio of 2. A power level that is changed by -3 dB is reduced to half the original power. A power level that is changed by $+3$ dB is doubled.
- 6 dB corresponds to a voltage ratio of 2. A voltage that is changed by -6 dB is reduced to half the original voltage. A voltage that is changed by $+6$ dB is doubled.
- 10 dB corresponds to a power ratio of 10. This is the only point where the dB value and the ratio value are the same (for power).

1.19 ABSOLUTE DECIBEL VALUES

Besides being useful for expressing ratios of powers or voltages, decibels can be used for specifying absolute voltages or powers. Either a power reference or voltage reference must be specified. For power calculations;

$$\text{dB (absolute)} = 10 \log \left(\frac{P}{P_{\text{REF}}} \right)$$

and for voltage calculations,

$$\text{dB (absolute)} = 20 \log \left(\frac{V}{V_{\text{REF}}} \right)$$

1.19.1 dBm

A convenient and often-used power reference for instrumentation use is the milliwatt (1 mW or 0.001 watt), which results in dBm:

$$\text{dBm} = 10 \log \left(\frac{P}{0.001} \right)$$

This expression is valid for any impedance or resistance value. As long as the impedance is known and specified, dBm can be computed using voltage. For a 50-ohm resistance, 1 mW of power corresponds to a voltage of 0.224 volts RMS:

$$P = \frac{V_{\text{RMS}}^2}{R}$$

$$V_{\text{RMS}} = \sqrt{P \cdot R} = \sqrt{0.001 \times 50} = 0.224$$

Using this voltage as the reference value in the decibel equation results in

$$\text{dBm}(50\Omega) = 20 \log \left(\frac{V_{\text{RMS}}}{0.224} \right)$$

Similarly, for 600 Ω and 75 Ω,

$$\text{dBm } (600\Omega) = 20 \log \left(\frac{V_{\text{RMS}}}{0.775} \right)$$

$$\text{dBm } (75\Omega) = 20 \log \left(\frac{V_{\text{RMS}}}{0.274} \right)$$

These equations are valid only for the specified impedance or resistance.

1.19.2 d/BV

Another natural reference to use in measurements is one volt. This results in the following:

$$\text{dBV} \quad 20 \log \left(\frac{V}{1} \right)$$

or simply

$$dBV = 20 \log(V)$$

This equation is valid for any impedance level (since it's reference is a voltage). Table 1-4 summarizes these equations and Figures 1-18 and 1-19 are plots of the basic decibel relationships. Table 1-5 is a table of voltage and power ratios along with their decibel values. The reader is encouraged to spend some time studying these figures to get a feel for how decibels work.

Example 1-7

A 0.5-volt RMS voltage is across a 50-Ω resistor. Express this value in dBV and dBm. What would the values be if the resistor were 75Ω?

50-Ω case

$dBV = 20 \log(0.5) = -6.02 \text{ dBV}$

$dBm (50\Omega) = 20 \log \left(\dfrac{0.5}{0.224} \right) = +6.97 \text{ dBm}$

TABLE 1-4 SUMMARY OF EQUATIONS RELATING TO DECIBEL CALCULATIONS. VOLTAGES (V) ARE RMS VOLTAGE, AND POWER (P) IS IN WATTS.

$dB = 10 \log \left(\dfrac{P_2}{P_1} \right)$

$dB = 20 \log \left(\dfrac{V_2}{V_1} \right)$

$dBm = 10 \log \left(\dfrac{P}{0.001} \right)$

$dBV = 20 \log(V)$

$dBm (50 \ \Omega) = 20 \log \left(\dfrac{V}{0.224} \right)$

$dBm (75 \ \Omega) = 20 \log \left(\dfrac{V}{0.274} \right)$

$dBm (600 \ \Omega) = 20 \log \left(\dfrac{V}{0.775} \right)$

$dBW = 10 \log(P)$

$dBf = 10 \log \left(\dfrac{P}{1 \times 10^{-15}} \right)$

$dB\mu V = 20 \log \left(\dfrac{V}{1 \times 10^{-6}} \right)$

$dBc = 10 \log \left(\dfrac{P}{P_{CARRIER}} \right)$

$dBc = 20 \log \left(\dfrac{V}{V_{CARRIER}} \right)$

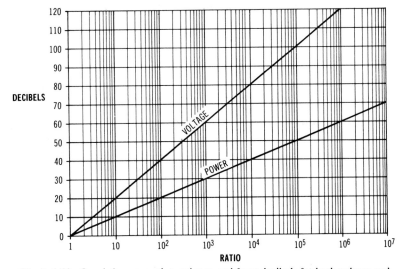

Figure 1-18 Graph for converting ratios to and from decibels for both voltage and power. Because of the logarithmic relationships, the plots are straight lines on log axes.

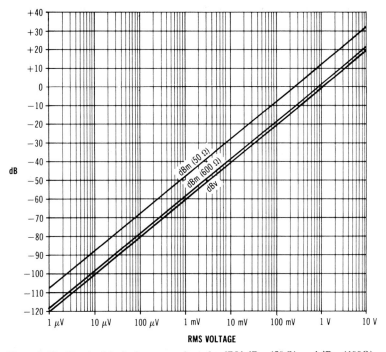

Figure 1-19 Graph of decibels versus voltage for dBV, dBm (50 Ω), and dBm (600Ω).

TABLE 1-5 TABLE OF DECIBEL VALUES FOR VOLTAGE RATIOS
AND POWER RATIOS.

Decibels	Power Ratio	Voltage Ratio
100	10,000,000,000	100,000
90	1,000,000,000	31,623
80	100,000,000	10,000
70	10,000,000	3,162
60	1,000,000	1,000
50	100,000	316.2
40	10,000	100.0
30	1,000	31.62
20	100	10.00
10	10	3.162
0	1	1.000
−10	0.1	0.3162
−20	0.01	0.1000
−30	0.001	0.03162
−40	0.0001	0.01000
−50	0.00001	0.003162
−60	0.000001	0.001000
−70	0.0000001	0.0003162
−80	0.00000001	0.0001000
−90	0.000000001	0.00003162
−100	0.0000000001	0.00001000

Decibels	Power Ratio	Voltage Ratio
10	10.0000	3.1623
9	7.9433	2.8184
8	6.3096	2.5119
7	5.0119	2.2387
6	3.9811	1.9953
5	3.1623	1.7783
4	2.5119	1.5849

75-Ω case

dBV is based on voltage so that value is the same as the 50-Ω case ($-$ 6.02 dBV).

dBm is referenced to 1 mW of power. We must either use the voltage equation for dBm (75Ω) or compute the power and use the power equation. We will compute the power.

$$P = \frac{V_{RMS}^2}{R} = \frac{0.5^2}{75} = 3.33 \text{ mW}$$

$$dBm = 10 \log\left(\frac{P}{0.001}\right) = 10 \log\left(\frac{0.00333}{0.001}\right)$$

$$dBm = 5.23 \text{ dBm}$$

TABLE 1-5 (Continued)

Decibels	Power Ratio	Voltage Ratio
3	1.9953	1.4125
2	1.5849	1.2589
1	1.2589	1.1220
0.9	1.2303	1.1092
0.8	1.2023	1.0965
0.7	1.1749	1.8039
0.6	1.1482	1.0715
0.5	1.1220	1.0593
0.4	1.0965	1.0471
0.3	1.0715	1.0351
0.2	1.0471	1.0233
0.1	1.0233	1.0116
0	1.0000	1.0000
−0.1	0.9772	0.9886
−0.2	0.9550	0.9772
−0.3	0.9333	0.9661
−0.4	0.9120	0.9550
−0.5	0.8913	0.9441
−0.6	0.8710	0.9333
−0.7	0.8511	0.9226
−0.8	0.8318	0.9120
−0.9	0.8128	0.9016
−1	0.7943	0.8913
−2	0.6310	0.7943
−3	0.5012	0.7079
−4	0.3981	0.6310
−5	0.3162	0.5623
−6	0.2512	0.5012
−7	0.1995	0.4467
−8	0.1585	0.3981
−9	0.1259	0.3548
−10	0.1000	0.3162

1.19.3 Other References

Other power reference values are 1 watt (dBW) and 1 femtowatt or 1×10^{-15} watt (dBf). A common voltage reference is the microvolt, resulting in dBµV. These values can be computed by the following equations:

$$dBW = 10 \log(P)$$

$$dBf = 10 \log\left(\frac{P}{1 \times 10^{-15}}\right)$$

$$dBuV = 20 \log\left(\frac{V}{1 \times 10^{-6}}\right)$$

Voltage or power measurements relative to a fixed signal (called the carrier) are expressed as dBc.

$$dBc = 10 \log\left(\frac{P}{P_{CARRIER}}\right)$$

or

$$dBc = 20 \log\left(\frac{V}{V_{CARRIER}}\right)$$

The reader will undoubtedly encounter other variations. The number of possible permutations on the lowly decibel is limited only by the imagination of engineers and technicians worldwide.[5]

1.20 MEASUREMENT ERROR

Because imperfections in any measuring instrument will disturb the circuit being tested by removing energy from that circuit (no matter how small), some amount of error will always be introduced. Another way of saying this is that connecting an instrument to a circuit changes that circuit and changes the voltage or current that is being measured. This error can be minimized by careful attention to the loading effect as discussed later.

In addition, errors internal to the instrument degrade the quality of the measurement. These errors fall into two main categories:

- *Accuracy*: the ability of the instrument to measure the true value to within some stated error specification.

- *Resolution*: the smallest change in value that an instrument can detect.

-

Suppose a voltage measuring instrument has an accuracy of ± 1 percent of the measured voltage and 3-digit resolution. If the measured voltage was 5 volts, then the accuracy of the instrument is 1 percent of 5 volts or ± 0.05 volts. The instrument could read anywhere from 4.95 to 5.05 volts with a resolution of 3 digits. The meter cannot, for instance, read 5.001 volts since that would require 4 digits of resolution.

If the instrument had 4-digit resolution (but 1 percent accuracy), then the reading could be anywhere from 4.950 to 5.050 volts (the same basic accuracy, but with more digits). Assume for the moment that both meters read exactly 5 volts. If the actual voltage changed to 5.001 volts, the 3-digit instrument would probably not register any change, but the 4-digit instrument would have enough resolving power to show that the voltage had indeed changed. (Actually, the 3-digit meter reading might change, but could not display 5.001 volts because the next highest possible reading is 5.01 volts.) The accuracy of the 4-digit instrument is

---

[5]It is rumored that some electrical engineers balance their checkbooks using dB$.

not any better, but it has finer resolution since it measures smaller changes. Neither meter guarantees that a voltage of 5.001 volts can be measured any more accurately than 1 percent.

The example given is a digital instrument, but the same concepts apply to analog instruments. Resolution is not usually specified in digits in the analog case, but all instruments have some fundamental limitation to their measurement resolution. Typically, the limitation is the physical size of the analog meter and its markings. For instance, an analog voltmeter with full scale equal to 10 volts will have difficulty detecting the difference between two voltages which differ by a few microvolts. Usually, an instrument provides more measurement resolution than measurement accuracy. This guarantees that the resolution will not limit the obtainable accuracy and allows detection of small changes in readings (even if those readings have some absolute error in them). The relative accuracy between small steps may be much better than the full-scale absolute accuracy.

1.21 THE LOADING EFFECT

In general, when two circuits are connected together, the voltages and currents in those circuits both change. One circuit is referred to as the source and the other circuit as the load. The source might be, for example, the output of an amplifier, transmitter, or signal generator. The corresponding load might be a speaker, antenna, or the input of a circuit. In the case where an electronic measurement is being made, the circuit under test is the source, and the measuring instrument is the load. The measuring instrument always has an effect on the circuit, but if this loading effect is small enough, it can be ignored.

Many source circuits can be represented by a simple circuit model called a THÉVENIN EQUIVALENT CIRCUIT. A Thévenin equivalent circuit is made up of a voltage source, V_S, and a series resistance, R_S (Figure 1-20). Thus, complex circuits can be simplified by replacing them with an equivalent circuit model. In the same way, many loads can be replaced conceptually with a circuit model consisting of a single

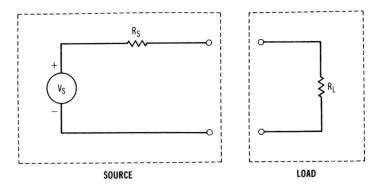

SOURCE   **LOAD**

Figure 1-20 A voltage source with internal resistance and a resistive load (Thévenin equivalent circuit).

resistance, R_L. (Resistive circuits will be considered initially in this discussion but later expanded to include AC impedances.)

V_S is also known as the OPEN-CIRCUIT VOLTAGE, since it is the voltage across the source circuit when no load is connected to it. This is easily proven by noting that no current can flow through R_S under open-circuit conditions. Therefore, there is no voltage drop across R_S and $V_L = V_S$.

1.22 THE VOLTAGE DIVIDER

When the load is connected to the source, the voltage across the source (and the load), V_L is no longer the open-circuit value, V_S. V_L is given by the voltage divider equation (named for the manner in which the total voltage in the circuit divides across R_S and R_L).

$$V_L = \frac{V_S R_L}{(R_S + R_L)}$$

Figure 1-21 shows a source and load connected in this manner. The resulting output voltage, V_L, as a function of the ratio R_L/R_S is plotted in Figure 1-22. If R_L, is very small compared to R_S, then V_L is also very small. For large values of R_L (compared to R_S), V_L approaches V_S.

1.23 MAXIMUM VOLTAGE TRANSFER

To get the maximum voltage out of a voltage source being loaded by some resistance, it is necessary to make the ratio R_L/R_S as large as possible. From a design point of view, this can be approached from two directions: make R_S small or make R_L large. Ideally, we could make $R_S = 0$ and make $R_L = $ infinity, resulting in $V_L = V_S$. In practice, this cannot be obtained, but can be approximated. Figure 1-22 shows that

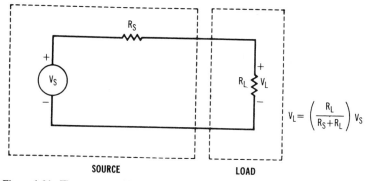

SOURCE   LOAD

Figure 1-21   The source and load are connected together. The voltage across the load is given by the voltage divider equation. The loading effect causes this voltage to be less than the open-circuit voltage of the source.

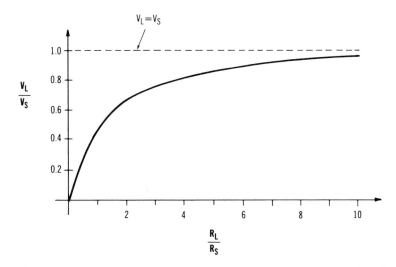

Figure 1-22 A plot of the output voltage due to the loading effect, as a function of the load resistance divided by the source resistance. The larger the load resistance, the better the voltage transfer.

making R_L 10 times larger than R_S results in a voltage that is 91 percent of the maximum attainable (V_S).

When making measurements, the source is the circuit under test and the load is the measuring instrument. We may not have control over the value of R_S (if it is part of the circuit under test) and our only recourse is choosing R_L (the resistance in our measuring instrument). Ideally, we would want the resistance of the instrument to be infinite, causing no loading effect. In reality, we will settle for an instrument which has an equivalent load resistance that is much larger than the equivalent resistance of the circuit.

1.24 MAXIMUM POWER TRANSFER

Sometimes the power delivered to the load resistor is more important than the voltage. Since power depends on both voltage and current ($P = V \times I$), maximum voltage does not guarantee maximum power. The power delivered to R_L can be determined as follows:

$$P = \frac{V_S \cdot R_L}{(R_L + R_S)} \frac{V_S}{(R_L + R_S)}$$

$$= \frac{V_S^2 R_L}{(R_L + R_S)^2}$$

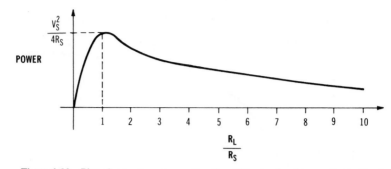

Figure 1-23. Plot of output power as a function of the load resistance divided by the source resistance. The output power is maximum when the two resistances are equal.

This relationship is plotted for varying values of R_L/R_S in Figure 1-23. For small values of R_L (relative to R_S), the power delivered to R_L is small, because the voltage across R_L is small. For large values of R_L (relative to R_S), the power delivered is also small, because the current through R_L is small. The power transferred is maximum when $R_L/R_S = 1$ or, equivalently, when R_L equals R_S.

$$R_L = R_S \quad \text{for maximum power}$$

In many electronic systems, maximum power transfer is desirable. Such systems are designed with all source resistances and load resistances being equal to maximize the power. Maximum power transfer is achieved while sacrificing maximum voltage transfer between the source and load.

1.25 IMPEDANCE

The previous discussion has assumed resistive circuits, but many circuit components are REACTIVE, exhibiting a phase shift between their voltage and current. This type of voltage and current relationship is commonly represented through the use of COMPLEX IMPEDANCE. The impedance of a device is defined as

$$Z = \frac{V_{0-P} \angle \theta_V}{I_{0-P} \angle \theta_I}$$

where

V_{0-P} = the AC voltage across the impedance
θ_V = the phase angle of the voltage
I_{0-P} = the AC current through the impedance
θ_I = the phase angle of the current.

Simplifying the equation,

$$Z = \frac{V_{0-P}}{I_{0-P}} \angle \theta_Z$$

where θ_Z is the phase angle of the impedance and is equal to $\theta_V - \theta_I$.

The preceding equations show the complex impedance in magnitude and phase format. Alternatively, the impedance can be expressed in a rectangular format.[6]

$$Z = R + j \cdot X$$

where

      R = the resistive component of impedance
      X = the reactive component of the impedance
      j = the square root of -1

For the purposes of maximum power transfer, when the source impedance and load impedance are not resistive, maximum power transfer occurs when the load impedance has the same magnitude as the source impedance, but with opposite phase angle.[7] For instance, if the source impedance were 50 ohms with an angle of +45 degrees, the load impedance should be 50 ohms with an angle of −45 degrees to maximize the power transfer. Mathematically, this can be stated as

$$Z_L = Z_S{}^*$$

where * indicates the complex conjugate.

A special case is when the phase angle of the source impedance is zero. In that case, the load impedance should have the same magnitude as the source impedance and an angle of zero (load impedance equals source impedance). Note that this is the same as the purely resistive case.

1.26 INSTRUMENT INPUTS

The characteristics of instrument input impedances vary quite a bit with each individual model, but in general they can be put into two categories: high impedance and system impedance.

1.36 High-Impedance Inputs

High-impedance inputs are designed to maximize the voltage transfer from the circuit under test to the measuring instrument by minimizing the loading effect. As outlined previously, this can be done by making the input impedance of the instrument much larger than the impedance of the circuit. Typical values for instrument input impedance are between 10kΩ and 1M Ω. For instruments used at high frequencies, the capacitance across the input becomes important and therefore is usually specified by the manufacturer.

---

[6]See Appendix D for information on rectangular and polar format numbers.

[7]For a derivation of this, see Irwin (1984).

1.26.2 System Impedance Inputs

Many electronic systems have a particular system impedance such as 50 ohms (Figure 1-24). If all inputs, outputs, cables, and loads in the system have the same resistive impedance, then, according to the previous discussion, maximum power is always transferred. At high frequencies (greater than about 30 MHz), stray capacitance and transmission line effects may make this the only type of system that is practical. The system impedance is often called the CHARACTERISTIC IMPEDANCE and is represented by the symbol Z_0.

At audio frequencies, a constant system impedance is not mandatory but is often used. It is sufficient for many applications to make all source circuits have a low impedance (less than 100 Ω) and all load circuits have a high impedance (greater than 1 kΩ). This results in near maximum voltage transfer (power transfer is less important here). Some audio systems do maintain, or at least specify for test purposes, a system impedance which is 600 Ω. This same impedance appears in telephone applications as well.

For radio frequency work, 50 Ω is by far the most common impedance. This impedance can be easily maintained despite stray capacitance, and 50-Ω cable is easily realizable. Such things as amateur and commercial radio transmitters, transmitting antennas, communications filters, and radio frequency test equipment all commonly have 50-Ω input and output impedances. Runner-up to 50 ohms in the radio frequency world is 75 ohms. This impedance is also used extensively at these frequencies, particularly in video-related applications such as cable TV. Other system impedances are used as special needs arise and will be encountered when making electronic measurements.

When measuring this type of system, most of the accessible points in the system expect to be loaded by the system impedance (Z_0). Because of this, many instruments are supplied with standard input impedance values (again, typically 50 ohms). Thus,

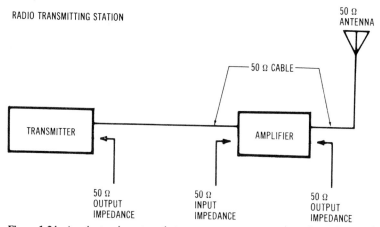

Figure 1-24 An electronic system that uses a common system impedance for maximum power transfer throughout.

the instrument can be connected into the system and act as a 50-Ω load while it's measuring.

1.27 BANDWIDTH

Instruments that measure AC waveforms generally have some maximum frequency above which the measurement accuracy is degraded. This frequency is the BANDWIDTH of the instrument and is usually defined as the frequency at which the instrument's response has decreased by 3 dB. (Other values such as 1 dB and 6 dB are also sometimes used.) A typical response is shown in Figure 1-25. Note that the response does not instantly stop at the 3-dB bandwidth. It begins sluggishly, decreasing at frequencies less than the bandwidth, and the response may still be usable for frequencies outside the bandwidth. Some instruments will have their bandwidth implied in their accuracy specification. Rather than give a specific 3-dB bandwidth, the accuracy specification will be given as valid only over a specified frequency range.

In general, to measure an AC waveform accurately, the instrument must have a bandwidth that exceeds the frequency content of the waveform. For a sine wave, the instrument bandwidth must be at least as large as the sine wave frequency. For sine wave frequencies equal to the 3-dB bandwidth, the measured value will, of course, be decreased by 3 dB, so an even wider bandwidth is desirable. The amount of margin necessary will vary depending on how quickly the response rolls off near the 3-dB point.

For sine wave measurements, the frequency response near the 3-dB bandwidth can be adjusted for by recording the response at these frequencies using a known input signal. Assuming the response is repeatable, the measurement error due to limited bandwidth can be used to adjust the measured value to obtain the actual value.

Instruments that measure only AC voltages (and not DC) have a response that rolls off at low frequencies as well as high frequencies. When using instruments of this

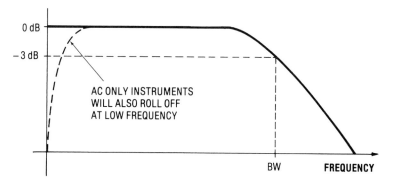

Figure 1-25 The frequency response of a typical measuring instrument rolls off at high frequencies. Some instruments that do not measure DC also roll off at low frequencies.

type (including many AC voltmeters), both the high-frequency and the low-frequency bandwidth limitations must be considered.

For waveforms other than sine waves, the harmonics must be considered. If the harmonics are outside the instrument bandwidth, their effect on the waveform will not be measured. This could be desirable if the harmonics outside some frequency range needed to be ignored. In general, their effect is usually desirable in the measurement. Waveforms with an infinite number of harmonics would require an instrument with infinite bandwidth to measure them. In reality, the higher harmonics have very little energy and can be ignored with proper regard to how much error this produces in the measurement.

1.28 RISE TIME

Ideally, waveforms such as square waves and pulses change voltage level instantaneously. In reality, waveforms take some time to make an abrupt change, depending on the bandwidth of the system and other circuit parameters. The amount of time it takes for a waveform to transition from one voltage to another is called the RISE TIME. (The rise time is normally measured at the 10 percent and 90 percent levels of the transition.) The bandwidth of a measuring instrument will limit the measured rise time of a pulse or square wave. For a typical instrument, the relationship between rise time and bandwidth is given by

$$t_{RISE} = \frac{0.35}{BW}$$

BW = 3-dB bandwidth (in hertz)

The validity of this relationship depends on the exact shape of the frequency response of the instrument (how fast it rolls off for frequencies above its bandwidth). It is exact for instruments with a single-pole roll-off[8] and is a good approximation for many instruments. The important point here is that the bandwidth, which is a frequency domain concept, will limit the measurement of the rise time, which is a time domain concept. The two characteristics of the instrument (time domain and frequency domain) are intertwined. Fast changes (small rise times) correspond to high-frequency content in a waveform, so if the high-frequency content is limited by the bandwidth of a system, then the rise time will be larger.

The instrument should have a rise time significantly smaller than the rise time being measured. A rise time measurement using an instrument with a rise time 2 times smaller than the one being measured will result in an error of about 10 percent. This error drops to 1 percent when the instrument rise time is 7 times shorter than the measured rise time.

The measured rise time can be computed from the signal rise time and the measuring instrument rise time.

---

[8]See Appendix D for a discussion of this type of frequency response and the derivation of the relationship between bandwidth and rise time.

$$t_{meas} = \sqrt{t_r^2 + t_{inst}^2}$$

where

t_{meas} = the measured rise time
t_r = the rise time of the signal
t_{inst} = the rise time of the measuring instrument.

1.29 BANDWIDTH LIMITATION ON SQUARE WAVE

As an example of how limited bandwidth can affect a waveform in the time domain, consider the square wave shown in Figure 1-26. The waveform is passed through a low-pass filter which has a frequency characteristic similar to Figure 1-25. If the

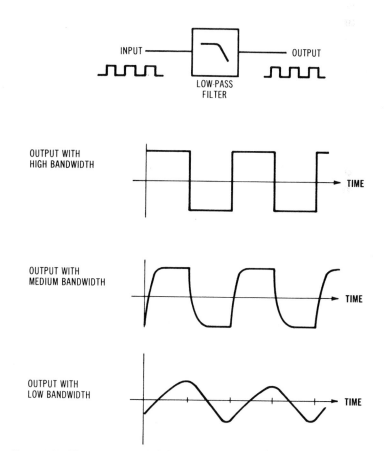

Figure 1-26 The effect of bandwidth on a square wave. With a very wide bandwidth, the square wave is undistorted; with a low bandwidth, the square wave is distorted.

bandwidth of the filter is very wide (compared to the fundamental frequency of the waveform), the square wave appears at the output undistorted. If the bandwidth is reduced, some of the harmonics are removed from the waveform. The output still looks like a square wave, but it has some imperfections that would cause a measurement error. For very limited bandwidth, the square wave barely appears at all and is very rounded due to the lack of high-frequency harmonics.

Example 1-8.

An electronic instrument is used to measure the voltage of a 2-kHz sine wave. What instrument bandwidth is required? What bandwidth would be required to measure a square wave having the same frequency? (Assume that any harmonic greater than 10 percent of the fundamental is to be included.)

Since a sine wave has only the fundamental frequency, the bandwidth of the instrument must be at least the frequency of the waveform. BW = 2 kHz. (In reality, we may want to choose a somewhat higher value since the instrument response is diminished by 3 dB at its bandwidth.)

From Table 1-3, the highest significant harmonic of a square wave (greater than 10 percent of the fundamental) is the 9th harmonic. Therefore, the bandwidth must be at least (and probably larger than)

$$BW = 9 \times 2 \text{ kHz} = 18 \text{ kHz}$$

Example 1-9.

A pulse with zero rise time is passed through a low-pass filter with a 3-dB bandwidth of 25 kHz. If the filter has a single-pole roll-off, what will the rise time be at the output?

Even though the rise time at the input is zero, due to the 25-kHz bandwidth limitation, the rise time at the output will be

$$t_{RISE} = \frac{0.35}{BW} = \frac{0.35}{25 \text{ kHz}} = 14 \text{ μsec}$$

Example 1-10.

A square wave with a rise time of 1 μsec is to be measured by an instrument. What 3-dB bandwidth is required in the measuring instrument to measure the rise time with 1 percent error?

For a 1 percent error in rise time, the measuring instrument must have a rise time that is 7 times less than the waveform's rise time.

For the instrument,

$$t_{RISE} = \frac{1 \text{ μsec}}{7} = 0.14 \text{ μsec}$$

$$BW = \frac{0.35}{t_{RISE}} = \frac{0.35}{0.14 \text{ μsec}} = 2.5 \text{ MHz}$$

1.30 DIGITAL SIGNALS

With the widespread use of microprocessor and other digital circuits, digital signals have become very common. The basic advantage (and simplicity) of a digital signal is that it has only two valid states: HIGH and LOW (Figure 1-27). The HIGH state is defined as any voltage greater than or equal to V_H, and the LOW state is defined as any voltage less than or equal to V_L. Any voltage between V_H and V_L is undefined. V_H and V_L are called the LOGIC THRESHOLDS.

A typical, well-behaved digital signal (a pulse train) is shown in Figure 1-28a. Whenever the voltage is V_{0-P}, the digital signal is interpreted as a HIGH. Whenever the voltage is zero, the digital signal is LOW. The digital signal in Figure 1-28b is not so well behaved. The signal starts out LOW, then enters the undefined region (greater than V_L but smaller than V_H), then becomes HIGH, then LOW, and HIGH once more. Notice that in the last LOW state, the voltage is not zero but increases a small amount. It does stay less than V_L, so it is still a LOW signal. Similarly, during the last HIGH state the voltage steps down a small amount, but stays above the high threshold.

Although the region between the two logic thresholds is undefined, it does not follow that it should never occur. First of all, to get from a LOW to a HIGH, the waveform must pass through the undefined area. Also, there are times when one digital gate goes into an "off" or high-impedance state while another gate drives the voltage HIGH or LOW. During this transition, the voltage could stay in the undefined area for a period of time. It is imperative that the digital signal settle to a valid logic level before the digital circuit following it uses the information. Otherwise, a logic error may occur.

Digital signals are mostly used to represent the binary numbers 0 and 1. If POSITIVE LOGIC is being used, then a HIGH state corresponds to a logical 1, and a LOW state corresponds to a logical 0. Although positive logic may appear to be the most obvious convention, NEGATIVE LOGIC is also sometimes used. With negative logic, the HIGH state represents a logical 0 and the LOW state represents a logical 1. Thus, the relationship between the voltage of the digital signal and the binary number which it represents varies depending on the logic convention used. Fortunately, the concepts associated with digital signals remain the same, but the instrument user may be left with a digital bookkeeping problem.

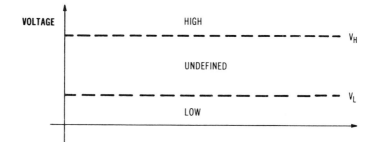

Figure 1-27. A digital signal must be one of two valid states, HIGH or LOW.

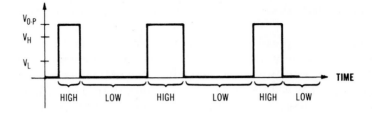

(A) A well behaved digital signal which is always a valid logic level.

(B) A digital signal that is undefined at one point.

Figure 1-28 Two digital signals.

1.31 LOGIC FAMILIES

There are several different technologies used to implement digital logic circuits. In general, they have unique values for the logic thresholds (V_H, V_L). To interpret a digital signal from a given logic technology or family, these thresholds must be known. Table 1-6 lists logic thresholds for the most popular logic families.

TABLE 1-6 LOGIC THRESHOLDS FOR THE MOST COMMON DIGITAL LOGIC FAMILIES. THE VALUES MAY VARY SLIGHTLY WITH DIFFERENT TYPES OF PARTS WITHIN THE SAME GENERAL FAMILY. ALL VALUES ARE IN VOLTS.

Logic Family	Supply Voltage	Input Threshold		Output Guaranteed		Output Typical	
		Low	High	Low	High	Low	High
TTL	5	0.8	2.0	0.4	2.4	0.2	3.5
CMOS	5	1.5	3.5	0.05	4.95	0	5.0
CMOS	10	3.0	7.0	0.05	9.95	0	10.0
CMOS	15	4.0	11.0	0.05	14.95	0	15.0
ECL	−5.2	−1.475	−1.105	−1.630	−0.980	−1.75	−0.90
ECL	5	3.525	3.895	3.370	4.020	3.25	4.10

TTL = Transistor Transistor Logic
CMOS = Complementary Metal Oxide Semiconductor
ECL = Emitter Coupled Logic

The logic thresholds are different depending on whether you are concerned with the input of a gate or the output of a gate. The V_L for the output of a gate is lower than the V_L for the input to a gate. Similarly, the V_H for the output is higher than V_H for the input. This slight amount of intentional overdesign guarantees that the output of a digital circuit will drive the input of the next circuit well past its required logic threshold. The difference between the input threshold and the output threshold is called the NOISE MARGIN, since the excess voltage is designed to overcome any electrical noise in the system. In general digital testing, the measurement instrumentation should usually be considered a digital input, and the logic levels associated with the input should be used.

REFERENCES

IRWIN, J. DAVID. *Basic Engineering Circuit Analysis.* New York: Macmillan Publishing Company, 1984.

OLIVER, BERNARD M., and JOHN M. CAGE. *Electronic Measurements and Instrumentation.* New York: McGraw-Hill Book Company, 1971.

SCHWARZ, STEVEN E., and WILLIAM G. OLDHAM. *Electrical Engineering: An Introduction.* New York: CBS College Publishing, 1984.

WEDLOCK, BRUCE D., and JAMES K. ROBERGE. *Electronic Components and Measurements.* Englewood Cliffs, NJ: Prentice-Hall, Inc., 1969.

WITTE, ROBERT A. *Spectrum and Network Measurements.* Englewood Cliffs, NJ: Prentice-Hall, Inc., 1991.

chapter two

Voltmeters, Ammeters, and Ohmmeters

Meters are the simplest, easiest-to-use instrument for measuring voltage, current, and resistance. A VOLTMETER measures voltage, an AMMETER measures current, and an OHMMETER measures resistance. Any meter might provide one or more of these functions. Convenience features such as autoranging (automatic selection of measurement range) combine with very-high-input impedance and digital display to provide uncomplicated meter measurements. For many applications, meter measurements have become as simple as selecting the meter function and connecting to the circuit under test. However, as the capability of a meter is stretched, more attention must be paid to the measurement details.

2.1 ANALOG AND DIGITAL METERS

Meters can be classified not only by the parameter that they measure but also by the type of display or readout used. Meter displays fall into two categories: Analog and digital.

2.1.1 Analog Meters

Analog meters generally use some type of electromechanical mechanism to cause a small arm to move depending on the voltage or current applied to the meter (Figure 2-1). A graduated measurement scale is imprinted behind the mechanism so that the moving arm points to the value of the meter reading. The actual mechanism involved is not critical; the important point is that analog meters provide a continuously varying readout, without any discrete jumps in meter reading. The mechanism does not have

Figure 2-1 An analog meter has a continuous scale. (Photo courtesy of B & K Precision—Dynascan Corporation.)

to be mechanical. It can be implemented using other technologies as long as the result is a continuous "analog" type of display.

Analog meters require care in observing meter readings. Different people watching the same measurement may disagree slightly on the correct interpretation of the meter reading. The resolution of the meter (the ability to measure small changes) depends on the physical layout of the meter, but generally it is no smaller than a few percent of the full-scale meter reading. Analog meters' biggest advantage when compared to digital meters is their ability to track changes in meter readings. Sometimes the ability to spot a trend in a meter reading (whether it is increasing or decreasing) is just as important as the absolute reading. For example, if a circuit is to be adjusted for maximum output voltage, an analog meter gives the user visual feedback for making the adjustment, and the output can be peaked very quickly.

2.1.2 Digital Meters

Digital meters, on the other hand, do not provide continuously variable meter readings. The meter reading is converted into decimal digits and is displayed as a number (Figure 2-2). This results in a reading that is very easy to interpret. Several different people viewing the same meter will record the same reading (except for, perhaps, cases where the number is changing due to electrical noise or measurement

Figure 2-2 A digital meter displays the meter reading as a decimal number.

drift). Digital meters are usually specified according to the number of digits in the display. A 3-digit voltmeter has three digits in its display. A 3 ½-digit voltmeter has three normal decimal digits plus a leading digit which has a limited range. Typically, the leading digit might only have allowable values of 0, 1, 2, or 3. Some meters are more restrictive, with a leading digit that can only be one or zero. The range of the display is specified by the manufacturer, often as a maximum "count." For instance, a 3200-count display on a 3-volt range can display voltages up to 3.200 volts.

Adjusting a circuit for maximum voltage is much more difficult with a digital meter. The user must read the number, decide whether the new reading is larger or smaller than the previous reading, and then tweak the adjustment in the proper direction. For meters with three or four digits, this can be difficult. Some digital meters have included a simple analoglike display in addition to the high-resolution digital display to provide the best of both technologies. The digital readout can be used for precise absolute measurements while the analog display can be used to adjust for maximum or minimum value.

It is important not to assume that the presence of many display digits automatically means very high accuracy. Often, a meter will provide more resolution than its accuracy specification supports. This ensures that the resolution of the meter does not limit its accuracy, but may mislead some users. The amount of accuracy versus resolution may vary with the different operating modes of the instrument, so careful examination of the specification sheet is in order.

2.2 DC VOLTMETERS

The IDEAL VOLTMETER (Figure 2-3) senses the voltage across its terminals without drawing any current because it has an infinite internal resistance. Therefore, it does not load the circuit under test (and, therefore, it is impossible to implement an ideal voltmeter). In other words, the ideal voltmeter looks like an open circuit to the

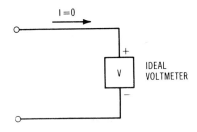

Figure 2-3 The ideal voltmeter senses the voltage across its terminals without drawing any current.

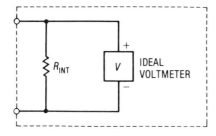

Figure 2-4 The real voltmeter has a finite internal resistance which draws some current from the circuit under test.

circuit under test. This is exactly what we want: a meter that can be connected to a circuit without disturbing it in any way. The internal implementation of the meter is not a concern, as the circuit model is valid for both analog and digital meters.

A more realistic model of a REAL VOLTMETER (Figure 2-4) includes the internal resistance of the meter. If the internal resistance of the meter is high enough compared with the resistance of the circuit under test, then there will be only minimal loading and the meter will approximate the ideal case. (See Chapter 1 for a discussion of the loading effect.)

Example 2-1.

A voltmeter with an internal resistance of 100 kΩ is used to measure a 10-volt source having an internal resistance of 2 kΩ (Figure 2-5). What is the error in the measurement due to loading?

If the voltmeter were ideal and did not load the circuit (Figure 2-6a), the measured voltage would simply be the value of the voltage source, 10 volts. But with a real voltmeter, the loading effect must be taken into account. Since the circuit is a voltage divider,

$$V_M = 10 \times \frac{100\text{k}\Omega}{(2\text{k}\Omega + 100\text{k}\Omega)} = 9.804 \text{ volts (real voltmeter)}$$

error = real − ideal = 9.804 − 10 = − 0.196 volts

The error is 2 percent of the ideal value.

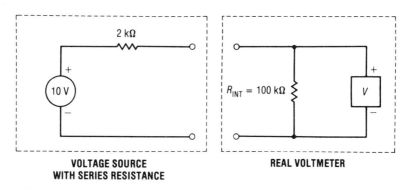

VOLTAGE SOURCE
WITH SERIES RESISTANCE

REAL VOLTMETER

Figure 2-5 A real voltmeter is used to measure the voltage source with an internal resistance.

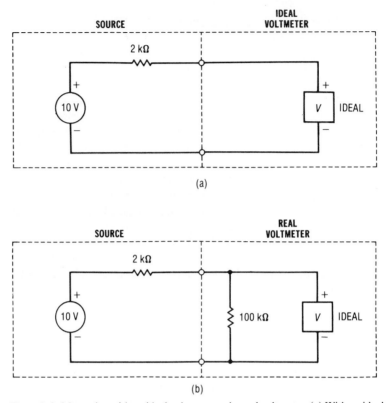

Figure 2-6 Measuring with an ideal voltmeter and a real voltmeter. (a) With an ideal meter, no current is drawn, and no voltage drop appears across the resistor. The meter reads 10 volts. (b) With a real meter, some current is drawn, and the loading effect reduces the measured voltage.

2.3 AC VOLTMETERS

The AC voltmeter is very similar to the DC voltmeter except that it measures the voltage of AC waveforms. (In some cases, it measures the DC and the AC portions of a waveform.) The same ideal voltmeter and real voltmeter circuit models shown in Figures 2-3 and 2-4 apply to the AC voltmeter. Again, the ideal voltmeter does not load the circuit under test, due to its infinite internal resistance. The loading effect of the real AC voltmeter is modeled by the resistance in Figure 2-4. A capacitor may be added in parallel with the resistance in the circuit model of the real meter to account for the input capacitance of the AC meter. (This is not a concern for DC meters since a capacitor acts like an open circuit for DC and does not load the circuit.)

2.3.1 Average-responding Meter

Unless otherwise specified, AC voltmeters are usually calibrated to read out in RMS volts. Not all meters, however, actually measure the RMS value of the signal.

Many low-cost meters use a circuit that responds to the average value of the waveform.[1] The meter readout is still calibrated to read RMS volts as long as the waveform is a sine wave. This is possible because the ratio of the average value and RMS value for a particular waveform can be determined. This is a perfectly valid technique as long as it is understood that the meter readout is correct only for sine waves. If, for example, a square wave were measured using an average-responding meter, the meter reading would be invalid. Some average-responding meters respond to the average value of the HALF-WAVE rectified waveform (instead of the full-wave rectified waveform). These meters will also be in error when used on anything but sine waves.

Some manufacturers supply correction factors with their average-responding meters such that other waveforms can be measured. This obviously requires that the shape of the waveform be known, since each waveform will have a different correction factor. If waveforms other than sine waves must be measured using an average-responding meter and the manufacturer has not supplied the correction factors, they may be determined experimentally by connecting the appropriate waveform with a known RMS voltage. This is a fairly reliable technique as long as the shape and frequency of the waveform are not varied.

2.3.2 Peak-reading Meters

Some AC voltmeters use a peak-detecting circuit that responds to the peak-to-peak value of the waveform. The meter display may be in terms of peak-to-peak volts or RMS volts. If the meter reads in peak-to-peak volts, the measured value should be accurate regardless of the shape of the waveform. However, if the meter reads RMS volts, then the reading is valid only for a sine wave. In a manner very similar to the average-responding meter, the peak-reading meter internally scales its measured voltage to obtain an RMS reading. This is done for only one waveform—the sine wave. So the peak-to-peak reading will be accurate independent of waveform, but the RMS reading is valid only for a sine wave.

2.3.3 True RMS Meters

Meters which respond to the actual RMS value of the waveform are called true RMS meters.[2] This simplifies the measurement, since the meter reads the correct RMS voltage regardless of the type of waveform. The root-mean-square function described in Chapter 1 is not simple to implement. Historically, these meters have been rather expensive to build, especially if a wide bandwidth was required. As integrated circuit technology has advanced in recent years, the price of true RMS has been decreasing dramatically and this feature is now appearing in low-cost meters.

---

[1] Here, average value refers to the full-wave rectified average value discussed in Chapter 1.

[2] The term "true" RMS is used to distinguish this type of meter from average-responding meters which are calibrated to read RMS voltage.

Even true RMS meters have some limitation on the shape of the waveform that can be measured. This limitation is usually specified as a maximum crest factor (peak-to-RMS ratio) that can be tolerated, perhaps with some specified accuracy. As a waveform's crest factor increases, a meter has to handle a larger peak voltage while measuring a relatively smaller RMS value.

2.3.4 Bandwidth

When making AC measurements, the bandwidth of the meter must be considered. Sometimes the bandwidth of the meter is specified directly, but more often the manufacturer provides an accuracy specification that depends on the frequency. Either way, the useful frequency range of the meter is defined.

For sine wave, the bandwidth must be at least as high as the frequency of the waveform. For other waveforms that contain harmonics, the harmonics must be included (see the discussion of bandwidth in Chapter 1).

2.3.5 AC and DC Coupling

For waveforms containing both DC and AC components, the AC voltage may be thought of as riding on top of the DC level. If an AC voltmeter is used to measure this type of voltage, the resulting meter reading will greatly depend on the design of the meter. Some manufacturers have included a coupling capacitor on the input of their voltmeters so that the meter is AC COUPLED. This AC coupling capacitor will block any DC voltage that is present but lets the AC portion of the waveform pass through to the meter. So if a meter is AC coupled, only the AC portion of the voltage is measured. (This concept of AC and DC coupling also applies to other instruments, particularly oscilloscopes.)

Some instruments do not have a coupling capacitor and respond to the DC (and AC) voltage present. This is referred to as DC COUPLING. It is important to understand how a particular meter behaves, if mixed DC and AC voltages are to be measured. Suppose the waveform in Figure 2-7 is to be measured by both types of

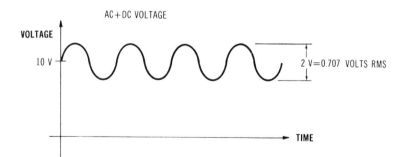

Figure 2-7 The measured value of a waveform containing both AC and DC will depend on the design of the meter. In this example, an AC coupled meter will read 0.707 volts RMS and a DC coupled meter will read 10.025 volts RMS.

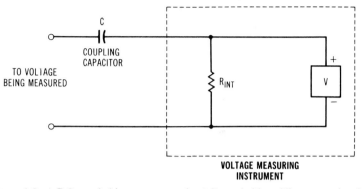

Figure 2-8 A DC coupled instrument can be AC coupled by adding a capacitor in series with the input.

meter. If the meter is AC coupled, then only the 2-volt peak-to-peak sine wave will be measured, resulting in $V_{RMS} = 0.707$ volts. On the other hand, if the meter is DC coupled, then both the AC and DC will be measured, resulting in a much higher reading. One might expect that the RMS reading in this case is just the sum of the DC and AC (RMS) voltages. This is not the case, since the RMS function of the voltmeter does not simply add the two voltages together. Assuming that the meter is a true RMS meter, it will read

$$V_{RMS} = \sqrt{2_{DC}^2 + V_{AC}^2} = \sqrt{10^2 + 0.707^2} = 10.025 \text{ volts}$$

If a meter is DC coupled, adding an external capacitor will make it AC coupled as shown in Figure 2-8. The capacitor will cause the meter response to roll off at low frequencies. The 3-dB point will be given by

$$f(3 \text{ dB}) = \frac{1}{2\pi \cdot R_{INT} \cdot C}$$

where R_{INT} is the internal resistance of the voltmeter. The value of the capacitor should be chosen such that the 3-dB frequency is about 10 times smaller than the frequency of the waveform being measured. This ensures that the frequency being measured is well within the bandwidth of the meter-capacitor combination.

2.4 RF PROBES

Since it is difficult to design an AC voltmeter that has very wide bandwidth, a RADIO FREQUENCY PROBE (RF probe) is often used to enable a DC voltmeter to make high-frequency AC measurements. An RF probe usually detects the peak of the AC waveform and converts it into a DC voltage that is then measured by a DC voltmeter. The probe is adjusted so that the DC voltmeter reads RMS volts, even though it is really responding to the peak value. Therefore, the probe is calibrated only for sine waves. Measuring any other waveform will result in errors (similar to the errors resulting from using a peak-reading AC meter). The RF probe will have some finite

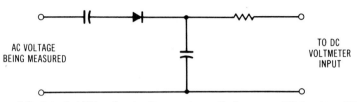

Figure 2-9 A typical RF probe circuit converts a radio frequency AC signal to a DC level readable by a DC voltmeter.

frequency range over which it will work (100 kHz to 500 MHz, for example). The frequency of the waveform being measured must, of course, fall within this bandwidth. Figure 2-9 shows a typical RF probe circuit.

Example 2-2.

The RMS value of three waveforms are to be measured: 1) a 60-Hz sine wave, 2) a 100-Hz triangle wave, and 3) a 2-MHz sine wave.

The following AC voltmeters are available:

A. An average-responding meter, with 20-Hz to 10-MHz bandwidth

B. A true RMS meter with 5-Hz to 100-KHz bandwidth. Which meters will accurately measure which waveforms?

Meter A can accurately measure only sine waves, since it is an average-responding meter (assuming that no correction factors are available). Both sine waves are well within the bandwidth of the meter, so waveforms 1 and 3 can be measured.

Meter B can measure any waveform shape, since it is a true RMS meter, but the 2-MHz sine wave is outside the bandwidth of the meter. So meter B can measure waveforms 1 and 2.

2.5 AMMETERS

The IDEAL AMMETER (Figure 2-10) senses the current going through it while maintaining zero volts across its terminals. This implies that the meter must have zero internal resistance. In other words, the meter acts like a short circuit when connected to the circuit under test. Recall from Chapter 1 that current measurements are made by breaking the circuit at the point of interest and inserting the instrument such that the current being measured flows through the meter. Since it is connected in series and has zero ohms, it does not affect the circuit under test. An ideal ammeter cannot be achieved in practice, but meters that have a very low internal resistance can come close.

A more realistic model of a REAL AMMETER (Figure 2-11) includes the internal resistance of the meter. If the internal resistance of the meter is small enough

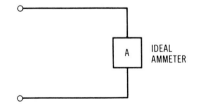

Figure 2-10 The ideal ammeter has zero internal resistance.

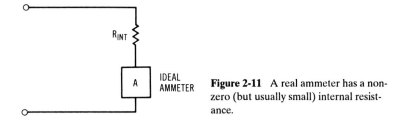

Figure 2-11 A real ammeter has a non-zero (but usually small) internal resistance.

compared with the resistance of the circuit under test, then there will be only a minimal effect on the circuit being measured, and the meter will approximate the ideal case.

Example 2-3.

An ammeter with an internal resistance of 100 ohms is used to measure the current shown in Figure 2-12a. What is the error in the measurement (due to the internal resistance of the ammeter)?

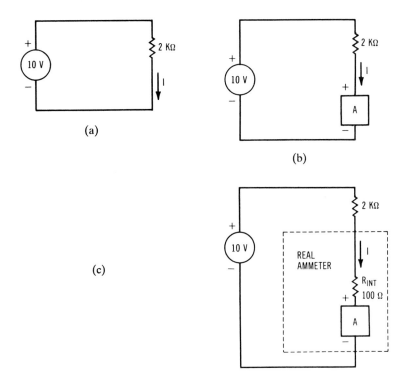

Figure 2-12 An example of current measurement using an ideal ammeter and a real ammeter. (a) Circuit to be measured. (b) Measurement using an ideal ammeter. (c) Measurement using a real ammeter.

If the ammeter were ideal and had zero internal resistance (Figure 2-12b), the measured current would simply be the value of the voltage source, 10 volts divided by the 2-kΩ resistance.

$$I = \frac{10\,V}{2\,k\Omega} = 5\,mA$$

But with a real ammeter, the internal resistance of the ammeter must be taken into account. Since the meter is in series with the circuit, the resistances add together.

$$I = \frac{10V}{(2\,k\Omega + 100)} = 4.76\,mA$$

error = real − ideal = 5 − 4.76 = 0.24 mA

The error is 4.8 percent of the ideal value.

2.6 AMMETER USED AS A VOLTMETER

An ammeter can be configured such that it measures voltage. Many meters, in fact, do use this technique internally to implement the voltmeter function while the actual metering mechanism responds to current. Figure 2-13 shows an ideal ammeter with a series resistance connected. (This series resistor should not be confused with the internal resistance of a real ammeter. If a real ammeter were being considered here, its internal resistance would add in series with R_S.) The current through the meter, I, is given by Ohm's law:

$$I = \frac{V_M}{R_S}$$

The current that the meter will read is proportional to the voltage being measured. For simplicity, consider the case where $R_S = 1\,k\Omega$.

$$I = \frac{V_M}{1k\Omega}$$

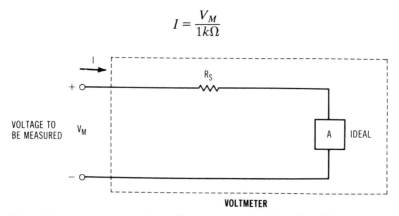

VOLTMETER

Figure 2-13 An ammeter can be configured to measure voltage by adding a resistor in series.

If the ammeter was originally calibrated to read milliamps, then in this configuration, it will display V_M in volts. The values in this example were chosen to illustrate the concept easily; in practice, other values for R_S may be used as long as the meter scale is chosen to display the voltage properly. Note that R_S is now the internal resistance of the simulated voltmeter. Since a large (ideally infinite) internal resistance is desired for a voltmeter, R_S is usually chosen to be fairly large and is limited by the sensitivity of the ammeter being used. (If R_S is chosen too large, then very little current will flow through the ammeter, requiring an ammeter capable of measuring very small currents.)

2.7 VOLTMETER USED AS AN AMMETER

A voltmeter can be configured such that it measures current. This technique can be used to make a current measurement even though only a voltmeter is available (or is more convenient). Figure 2-14 shows an ideal voltmeter connected in parallel with R_P. (This parallel resistor should not be confused with the internal resistance of a real voltmeter. If a real voltmeter were shown instead of an ideal voltmeter, its internal resistance would add in parallel with R_P.) Since the ideal voltmeter has no current flowing through it, all the current, I_M, must be flowing through R_P. Therefore, the voltage across the ideal voltmeter is

$$V = I_M \times R_P$$

The voltage displayed by the voltmeter is proportional to the current being measured. Suppose that R_P is equal to one ohm; then one amp of current through the new ammeter will correspond to a one-volt reading on the voltmeter. Other values can be used for R_P as long as the scale of the voltmeter is modified to provide the correct current reading. R_P is the internal resistance of the simulated ammeter, so it is desirable to make R_P as small as possible. The value of R_P will be limited by the sensitivity of the voltmeter and the amount of current being measured. For very small values of R_P, very little voltage will appear across it, requiring a very sensitive voltmeter.

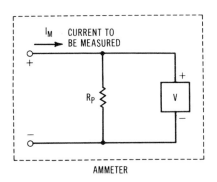

Figure 2-14 A voltmeter can be used to measure current by adding a resistor in parallel.

2.8 CURRENT-SENSE RESISTOR

Another interpretation of Figure 2-14 is that R_P is being used as a CURRENT-SENSE RESISTOR. Again, R_P is chosen to be very small so that the circuit being measured will not be disturbed. For example, suppose the current, I_M, in Figure 2-15a is to be measured. Figure 2-15b shows a small 1-Ω current-sense resistor placed in series where the current is to be measured. Since the 1-Ω resistor is very small compared to the 20-kΩ resistor, it will not affect the value of I_M significantly. It does, however, provide a handy place to connect a voltmeter or other voltage measuring instrument so that the current, I_M, can be determined. The current is calculated using the measured voltage (across the current-sense resistor) divided by the value of the current-sense resistor. In this particular example, the voltage across the 20-kΩ resistor could have just as easily been measured and used to determine the current. In many circuits, a resistor is either not available or not conveniently located for current measurement. In those cases, a small current-sense resistor can be added without affecting the operation of the circuit.

Example 2-4.

A current-sense resistor is to be used to measure a current that varies between 4 and 6 mA. If a voltmeter with full scale equal to 1 volt is used, what value of resistor will provide maximum sensitivity?

For maximum sensitivity, the voltmeter should read full scale when the current is 6 mA.

$$R = \frac{V}{I} = \frac{1}{0.006} = 166.7 \ \Omega$$

The resulting voltmeter reading in volts must be multiplied by 1/166.7 = 0.006 to determine the current in amps. (Multiply the voltage by 6 to get the current in milliamps.)

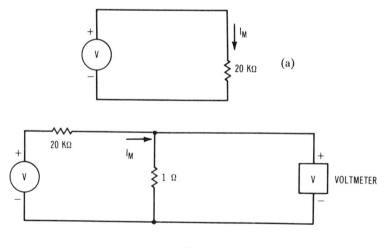

(b)

Figure 2-15 The use of a current-sense resistor is shown. (a) The current to be measured flows through the 20-kΩ resistor. (b) A small current-sense resistor is inserted in series and the voltage across it is measured.

The circuit under test should be evaluated to see if the current-sense resistor will disturb it significantly.

2.9 AC AMMETER

The AC ammeter is very similar to the DC ammeter except that it measures the current of AC (not DC) waveforms. The same ideal ammeter and real ammeter circuit models shown in Figure 2-10 and Figure 2-11 apply to the AC ammeter. Again, the ideal ammeter does not affect the circuit under test, due to its zero internal resistance, but a real ammeter will have some small resistance.

All the considerations discussed under AC voltmeters (average versus true RMS responding meters, bandwidth, AC versus DC coupling) apply to AC ammeters as well. The reader simply needs mentally to substitute "current" for every occurrence of "voltage."

2.10 OHMMETERS

An ohmmeter measures the resistance of the device or circuit connected to its input. There are many different meter configurations which can be used to implement an ohmmeter. Several of the more common configurations are described later in this chapter. It is usually not necessary to know the internal operation of the ohmmeter as long as the operating manual is followed correctly. What is important is that certain general principles be understood when making resistance measurements.

Since resistance measurements are derived from voltage and current measurements, ohmmeters supply their own stimulus (voltage or current) to the device being measured. This allows individual resistors to be measured without being part of a functioning circuit. This also means that if the resistance is part of a circuit, that circuit must have other sources of voltage (or current) removed from it. Power supplies and batteries, must be turned off or disconnected from the circuit. If not, currents or voltages induced will cause the ohmmeter reading to be incorrect and, if large enough, may damage the meter. (Compare this to voltage and current measurements, where the circuit *must* be powered up to get any meaningful readings.)

Ohmmeters operate using DC voltages and currents; therefore, the DC resistance of the device is measured (not AC impedance). Attempts to measure AC impedance with an ohmmeter result in inaccurate and frustrating readings. The classic example is the person who tries to verify the impedance of a loudspeaker, specified as 8 ohms, using an ohmmeter. The ohmmeter will actually measure the DC resistance of the speaker, which is generally just the resistance present in the voice coil winding. This reading could be most any value and does not indicate the AC impedance of the speaker. Also, measurements on components other than resistors such as diodes, transistors, and capacitors, etcetera may be misleading (although diode measurements are discussed later).

A zero resistance adjustment is often provided in an ohmmeter. This adjustment compensates for some of the measurement drift internal to the meter as well as things external to the meter such as test lead resistance. The usual procedure is to short the test leads together and adjust the meter until it reads zero. This procedure compensates for the resistance of the test leads, removing their effect from the measurement. If the test leads are removed or changed for a particular measurement, the meter should be zeroed again. In some meters it is also necessary to zero the meter on every different range setting. Instead of a zero adjustment some meters have an infinity adjustment. In that case, the test leads are held apart (open circuited), and the meter is adjusted to read infinity. Other meters have both zero and infinity adjustments which must be adjusted separately.

Resistance exists between two points, very similar to voltage. When an ohmmeter is connected to two points (for example, two ends of a resistor being measured), all the circuit paths from one point to the other will be measured. If there are two resistors in parallel between the two points, the parallel combination will be measured. If there are two resistors in series between the two measurement points, then the series combination will be measured. This must be carefully considered when measuring resistors that are part of a circuit. Simply connecting an ohmmeter across a resistor in a circuit does not necessarily measure only that resistor since there may be other components connected in parallel. Either these components must be accounted for in the resulting measurement or one end of the resistor being measured must be disconnected to guarantee that no other device is connected to it.

The human body has a resistance that varies considerably and affects high-resistance measurements. The conducting part of the test leads should not be touched by the user when making resistance measurements above about 10 kΩ; otherwise, body resistance may affect the measurement by appearing in parallel with the resistor being measured.

2.11 VOLTMETER-AMMETER METHOD

One obvious way to measure resistance is to connect a voltmeter and ammeter as shown in Figure 2-16. A voltage source and resistor, V_S and R_S, cause a current to flow through the resistance being measured (R_X). The current through and the voltage across R_X are measured by the voltmeter and ammeter and the value of R_X can be computed using Ohm's law, $R_X = V_M/I_M$.

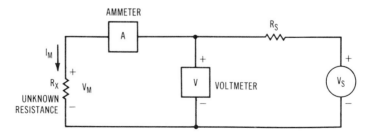

Figure 2-16 The voltmeter-ammeter method measures the voltage and current across the unknown resistance. The value of the resistance is then computed using Ohm's law.

This method is useful for making in-circuit tests when a voltmeter and a ammeter are available. In such a case, V_S and R_S are not used since the circuit operation presumably causes some current to flow through R_X, allowing the measurement to be performed. This method is not normally used in ohmmeters since it requires both an ammeter and a voltmeter to be implemented inside the ohmmeter.

2.12 SERIES OHMMETER

An ohmmeter implementation requiring only one meter is shown in Figure 2-17. R_X is the resistance being measured, while R_1 and V_S are known values internal to the ohmmeter. The resulting current through R_X can be computed as follows:

$$I_X = \frac{V_S}{R_1 + R_X}$$

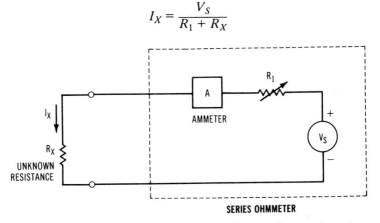

Figure 2-17 The series ohmmeter measures the unknown resistance using only one ammeter.

Note that I_X is *not* directly proportional to the value of R_X. Therefore, since the ammeter reading is used to indicate resistance, the ammeter does not have a simple linear scale but looks something like Figure 2-18 (assuming an analog meter is used). A full-scale current reading indicates zero resistance, while a zero current reading indicates infinite resistance. (The previous statement can be confirmed with a quick review of the circuit in Figure 2-17.)

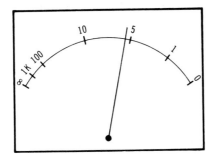

Figure 2-18 The meter scale for the series ohmmeter. Notice that it is not a simple linear scale.

R_1 is typically made variable to allow for precise calibration at the $R_X = 0$ point. Normally, the test leads of the ohmmeter are shorted together, and R_1 is adjusted until the meter reads zero ohms. This adjustment compensates for changes in V_S (which may be a battery) and for test lead resistance.

2.13 CURRENT-SOURCE METHOD

Another method used to implement an ohmmeter is to pass a known current through the unknown resistance and measure the resulting voltage (Figure 2-19). The current source, I_S, causes a fixed current to flow,[3] all of which passes through the R_X (since an ideal voltmeter has infinite resistance). Since $V_M = I_S R_X$, the measured voltage is directly proportional to the unknown resistance, so this method results in a linear relationship between the meter reading and the unknown resistance.

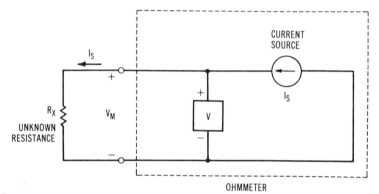

Figure 2-19 The current-source method for implementing an ohmmeter passes a known current through the resistor being measured. The voltmeter is then used to measure the voltage across the resistor, which is proportional to the resistance.

1.14 4-WIRE OHMS MEASUREMENTS

When measuring very small resistances, the resistance of the meter test leads may introduce a significant error. Figure 2-20a shows the circuit model of a current-source method ohmmeter with the resistance of the test leads included. The current I_S flows through these lead resistances and produces a voltage across each of them. Since the voltmeter will include these voltages in its measurement, the lead resistances add to the unknown resistance, R_X, producing an error. Since high-quality test leads have resistances of less than one ohm, this is only a problem when measuring small resistances.

The 4-WIRE OHMS technique removes this error by using two additional test leads. Figure 2-20b shows the test configuration for the 4-wire ohms measurement.

---

[3]If the concept of a current source is bothersome, consider it to be adjustable voltage source which always adjusts itself to the voltage required to cause the desired, fixed current to flow.

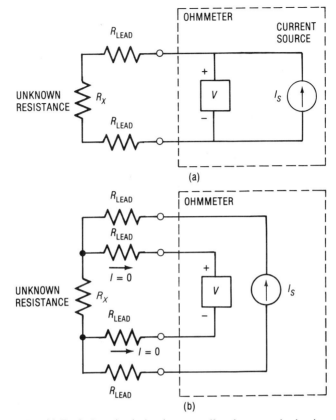

Figure 2-20 (a) Equivalent circuit showing error effect due to test lead resistance. (b) Equivalent circuit for the 4-wire ohms' measurement with test lead error eliminated.

The current I_S flows through a separate set of leads to the unknown resistance. These leads also have some finite resistance in them, but the current source will maintain a constant current to the unknown resistance. The other two test leads are used by the voltmeter to monitor the voltage across the unknown resistance. But since there is no current flowing through these two test leads, there is no voltage drop across them. Therefore, the voltmeter can accurately sense the voltage across R_X and since the current through the resistor is known, the resistance can be determined. The key to the 4-wire ohms technique is keeping the current out of the voltmeter leads. (The internal resistance of the voltmeter must be very high.)

2.15 MULTIMETERS

The most popular type of meter, available in both analog and digital form, is the MULTIMETER. A multimeter is simply a voltmeter, ammeter, and ohmmeter combined into one instrument. Internally, either an ammeter or voltmeter is implemented, which is then used to perform the other functions. Other names given to

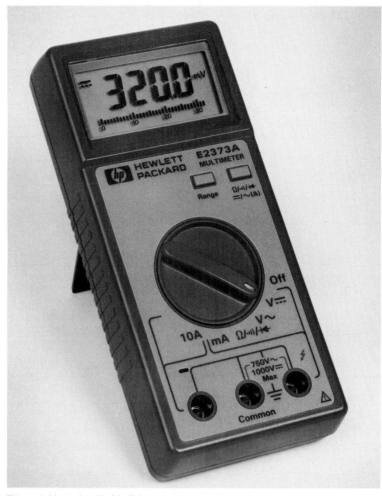

Figure 2-21 A handheld digital multimeter. (Photo courtesy of Hewlett-Packard Company.)

the multimeter are VOLT-OHM-MILLIAMMETER or VOM (usually an analog meter) and DIGITAL MULTIMETER or DMM (always a digital meter). An example of a handheld digital multimeter is shown in Figure 2-21, and a summary of its specifications is listed in Table 2-1.

2.16 METER RANGE

The techniques discussed for implementing voltmeters, ammeters, and ohmmeters are useful over a limited measurement range. Circuit techniques are used to expand or contract this measurement range to supply the user with different measurement settings. Range selection may be implemented by switching different resistor values

TABLE 2-1 ABBREVIATED SPECIFICATIONS OF A HANDHELD DIGITAL MULTIMETER.

Description: Handheld Digital Multimeter

DC volts
    Ranges: 300 mV, 3 V, 30 V, 300 V, 1000 V
    Accuracy: ±(0.7% of reading + 1 count)
    Input resistance: 10 MΩ minimum (depends on range)
AC volts
    Ranges: 3 V, 30 V, 300 V, and 750 V
    Accuracy: ±(1.2% of reading + 4 counts); maximum frequency, 500 Hz
DC current
    Ranges: 300 µA, 3 mA, 30 mA, 300 mA, 10 A
    Accuracy: ±(1.5% of reading + 5 counts)
Resistance
    Ranges: 300 Ω, 3 kΩ, 30 kΩ, 300 kΩ, 3 MΩ, 30 MΩ
    Accuracy: ±(0.7% of reading + 1 count); depends on range

into the metering circuit or by switching in an amplifier at the input of the meter. A multimeter functioning as a voltmeter, for example, might have 100-mV, 1-V, 10-V, and 100-V ranges. For maximum accuracy, the lowest range larger than the voltage to be measured is used. A larger range will work, but with reduced accuracy. A lower range will cause the meter to be overloaded, possibly damaging it. If the approximate value of the voltage being measured is unknown, then starting with the maximum range is recommended.

Meters with an autorange function automatically choose the best range for a given measurement. Autoranging meters essentially duplicate the actions of an experienced instrument user. If the meter is currently being overloaded, a higher range is selected. If the measured value is small enough that a lower range could be used (with greater accuracy), then a lower range is selected. Autoranging meters usually let the user disable the autorange feature and select the range directly. This is convenient when a particular range is required or the correct range is already known. Since it takes time for the meter to autorange, repetitive measurements near the same value may proceed faster if the range is selected directly.

2.17 CONTINUITY INDICATOR

Other features may be included along with the basic voltage, current, and resistance metering. When tracing out wires in a circuit, an ohmmeter can be used to determine whether or not a connection exists between two points. If the resistance value is very low (typically less than a few ohms), there is circuit continuity. If the resistance value is high, then no connection exists. Since the actual resistance measured is not all that critical for many continuity checks, some meters have an audible indicator that can be set to beep when the resistance is less than some value (say, 10 ohms). Thus, the user does not have to turn and look at the meter, but instead can focus on the circuit being tested.

2.18 DIODE TEST

Some meters provide a special diode test mode. A known current is forced through the diode, and the forward voltage drop across the diode is displayed. The current value is typically between 0.5 and 1 mA, large enough to switch most diodes on, but small enough so that sensitive diodes will not be damaged. Since a diode conducts in only one direction, proper polarity must be maintained when making the measurement. In some meters, the diode test mode is just a particular range of the ohmmeter function set up so the display reads out the voltage drop. Diode leakage (the flow of current in the reverse direction) may be tested by reversing the polarity of the leads.

2.19 SPECIFICATIONS

The exact method of specifying meter performance varies with both instrument model and manufacturer, but generalities do exist. Voltmeter, ammeter, and ohmmeter accuracy are usually specified as a percentage of full scale (i.e., percentage of the range) or as a percentage of the reading or both. Digital meters usually add to this an uncertainty of several digital counts.

Each different function of a meter (AC voltmeter, DC voltmeter, ohmmeter, etc.) as well as each range of each function may have separate specifications. It is important to examine the manufacturer's specifications carefully.

Example 2-5.

A 3 $\frac{1}{2}$-digit digital multimeter used on the 20-V range (maximum reading is 19.99 V) has an accuracy specification of $\pm(0.75\%$ of reading + 2 counts). What is the maximum error if the input voltage is 12 V?

One count is the smallest change that the meter can display (resolution), in this case, 0.01 V. (With 3 $\frac{1}{2}$ digits, the meter can read 12.00, 12.01, 12.02, etc.)

error = $\pm(0.75\% \times 12$ V$) + (2 \times 0.01)$

error = $\pm(0.09 + 0.02)$

error = ± 0.11 V

So the meter reading could be as low as 11.89 or as high as 12.11 V. (This analysis does not include other sources of error such as the loading effect described in Chapter 1.)

There are a variety of multimeters available in the marketplace, with a corresponding wide range of measurement capability as well as price. Generally, increased accuracy, resolution, and features cause a corresponding increase in cost. On the low end are small, portable analog VOMs. Also small and portable, but with improved accuracy and a digital display, is a compact digital multimeter such as the model shown in Figure 2-21. Figure 2-22 shows a digital multimeter with true RMS capability intended for bench use. Its specifications are listed in Table 2-2.

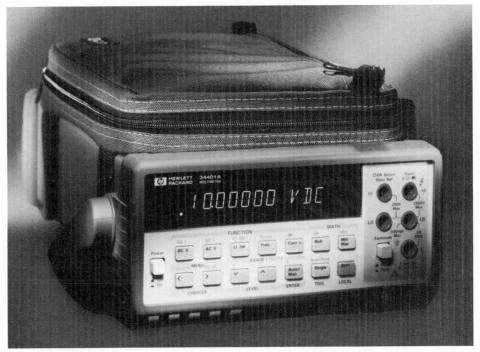

Figure 2-22. A bench top digital multimeter with true RMS detection. (Photo courtesy of Hewlett-Packard Company.)

TABLE 2-2 ABBREVIATED SPECIFICATIONS OF A BENCH DIGITAL MULTIMETER.

Description: 6 ½-Digit True RMS Autoranging Digital Multimeter

DC volts
  Ranges: 0.1 V, 1 V, 10 V, 100 V, 1000 V
  Accuracy: ±(0.005% of input + 0.0035% of range); depends on range
  Input resistance: 10 MΩ minimum
AC volts
  Ranges: 0.1 V, 1 V, 10 V, 100 V, 750 V
  Accuracy: ±(0.06% of input + 0.03% of range), 10 Hz to 20 kHz; depends on range, other
    frequencies also specified
DC current
  Ranges: 0.01 A, 0.1 A, 1 A, 3 A
  Accuracy: ±(0.1% of input + 0.01% of range); depends on range
Resistance
  Ranges: 100 Ω, 1 kΩ, 10 kΩ, 100 kΩ, 1 MΩ, and 10 MΩ
  Accuracy: ±(0.01% of input + 0.001% of range); depends on range

REFERENCES

GOTTLIEB, IRVING M. *Basic Electronic Test Procedures*, 2nd ed. Blue Ridge Summit, PA: Tab Books, Inc., 1985.

OLIVER, BERNARD M., and JOHN M. CAGE. *Electronic Measurements and Instrumentation.* New York: McGraw-Hill Book Company, 1971.

WEDLOCK, BRUCE D., and JAMES K. ROBERGE. *Electronic Components and Measurements.* Englewood Cliffs, NJ: Prentice-Hall, Inc., 1969.

chapter three

Signal Sources

Most of the instruments discussed in this book measure an electrical parameter, usually voltage or current. In this chapter, we will discuss a class of instruments that do not measure anything—at least not by themselves. Instead they provide the signals which stimulate the circuit or device being tested, so that other instruments may measure the electrical parameters of interest.

We will use the term SIGNAL SOURCE to mean an instrument that supplies a known signal (usually a voltage waveform) to a circuit being tested. The characteristics of that signal will vary depending on the particular type of signal source. Some sources produce only a single-frequency sine wave. Other, more complicated instruments provide signals such as triangle waves, square waves, and pulses and can sweep the frequency of the waveform as well as modulate the signal. Although various distinct categories of signal sources will be discussed, many instruments have characteristics which transcend several categories.

3.1 CIRCUIT MODEL

Although the actual circuitry of a signal source may be very complex, its electrical characteristics can be modeled with a simple circuit (Figure 3-1). V_S is the open circuit voltage of the source, and R_S is the output resistance. V_S is not limited to a simple sine wave, but can be any waveform that a particular signal source is capable of producing. This includes square and triangle waves, pulse trains and modulated signals. The output resistance is typically one of several common values: 50, 75, or 600 ohms. Most source specifications assume that the source is connected to an impedance that is equal to the source output resistance.

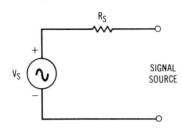

Figure 3-1 The circuit model for a signal
source is a voltage source with a series resistance.

Example 3-1.

A sine wave source with a 50-Ω output resistance is supplying 3 volts RMS across a 50-Ω
load resistor. How much power will be supplied to the load if the 50-Ω resistor is replaced
by a 75-Ω resistor?

Figure 3-2a shows the circuit model for the source connected to a 50-Ω load. Since $V_L =$
$V_S(50)/(50 + 50)$ by the voltage divider equation,

$$V_S = 2V_L = 6 \text{ V RMS}$$

Figure 3-2b shows the situation with the 75-Ω load resistor connected. In this case, $V_L =$
$6(75)/(75 + 50) = 3.6$ V RMS.

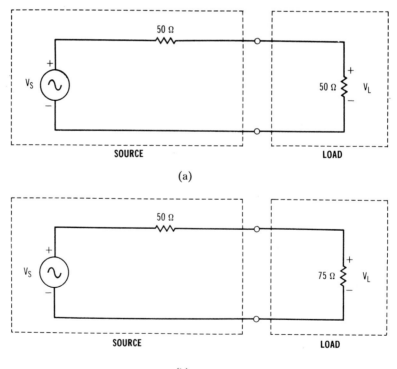

Figure 3-2 Circuit drawings for Example 3-1. (a) A signal source with a 50-Ω load
resistor. (b) The same signal source with a 75-Ω load resistor.

The power delivered to the load is

$$P = \frac{V_L^2}{R} = \frac{3.6^2}{75} = 0.173 \text{ watts}$$

The power can be expressed in dBm, $P = 10 \log(0.173/0.001) = 22.4$ dBm.

3.2 FLOATING AND GROUNDED OUTPUTS

The output of a signal source may be floating (Figure 3-3a) or grounded (Figure 3-3b). If the output is floating, then neither of the two output terminals is connected to the instrument's chassis ground. If the output is grounded, then one of the output terminals is connected to the instrument chassis ground which in turn is connected to the power line safety ground (via the ground connection on the three-terminal AC power plug). Some instruments offer a floating output with a switch or jumper built in to allow convenient grounding of the output if desired.

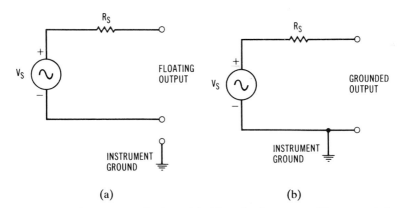

Figure 3-3. Signal source circuit model. (a) A floating output. (b) A grounded output.

Floating outputs are more versatile, but are more difficult to design and manufacture, especially at high frequencies. Many circuits have one side of their input connected to ground anyway, so a grounded output source can often be used with no problem. In some cases, however, it is desirable to have neither side of the signal source connected to ground. For example, the source may be driving a circuit at a point which has both input connections at several volts above ground. If a grounded source were connected, one of these input points would then be grounded and the operation of the circuit would be disturbed.

3.3 IMPERFECTIONS IN SIGNAL SOURCES

Signal sources do not produce absolutely perfect, undistorted signals, but instead a variety of imperfections may be present. Different applications require different levels of performance in the signal source, so the user must understand these imperfections and how they relate to the measurement. A pure, undistorted sine wave would appear in the frequency domain as shown in Figure 3-4a. Recall from Chapter 1 that a sine wave has energy at only one precise frequency with no harmonics or other frequency components. This perfect sine wave does not vary in amplitude or in frequency.

3.3.1 Frequency Accuracy

The FREQUENCY ACCURACY of a signal source determines how precisely the actual frequency of the waveform matches the frequency setting on the front panel of the instrument. It is usually specified in percent. A frequency error shows up in the frequency domain as a horizontal shift of the sine wave's vertical line (Figure 3-4b).

3.3.2 Frequency Stability

The sine wave from a signal source may vary somewhat in frequency depending on the source's FREQUENCY STABILITY. This results in a frequency error that varies with time. In the frequency domain, one can imagine an unstable signal wandering slightly back and forth along the frequency axis. This wandering is typically rather slow, but in extreme cases, a source could be so unstable that the resulting waveform has noticeable frequency modulation sidebands. There is a fine and arbitrary distinction between frequency instability and actual frequency modulation. Usually, a slow frequency variation (less than 1 Hz) is classified as frequency instability, and a fast variation is classified as RESIDUAL FREQUENCY MODULATION. Frequency stability is usually specified in percent or parts per million (ppm) and may be valid only after a specified warm-up time.

   Frequency stability and frequency accuracy are related but not the same. If a source had perfect frequency stability, but poor frequency accuracy, then the frequency would remain constant, but at the wrong value. The frequency inaccuracy of such a source could be compensated for by tuning it to the proper frequency while monitoring it with a more accurate instrument, such as a frequency counter. On the other hand, a source with poor frequency stability cannot achieve good frequency accuracy, at least not for any length of time. Such a source could be set to the correct frequency for an instant but would quickly drift away from it.

3.3.3 Amplitude Accuracy

AMPLITUDE ACCURACY specifies how precisely the actual amplitude of the waveform matches the instrument's control setting. An amplitude error in the frequency domain will result in a vertical line at the proper frequency, but with an incorrect height (amplitude) as shown in Figure 3-4c.

   Some instruments do not have a specified amplitude accuracy, but have an AMPLITUDE FLATNESS specification. Amplitude flatness is a measure of how much the output amplitude varies over a specific frequency range, usually the maxi-

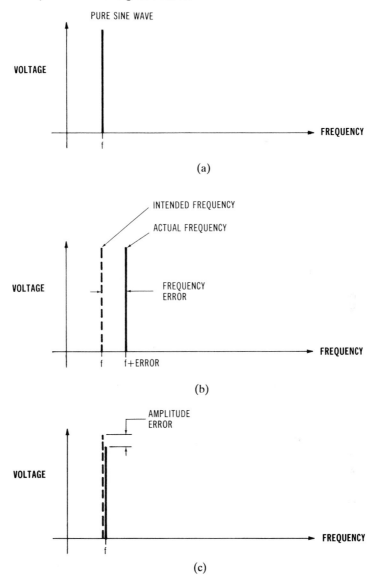

Figure 3-4 Signal source characteristics in the frequency domain. (a) A pure sine wave appears as a single-frequency component. (b) A frequency error shows up as a shift along the horizontal axis. (c) An amplitude error causes a change in the height of the spectral line. (d) Harmonics are additional frequency components at multiples of the original frequency. (e) Spurious responses are unwanted frequency components at other nonharmonic frequencies. (f) Close-in sidebands can be caused by residual amplitude or frequency or phase modulation.

mum range of the instrument. Once the user sets the source's amplitude (perhaps by monitoring it with another instrument), the amplitude will remain within the limits of the flatness specification even though the frequency is changed.

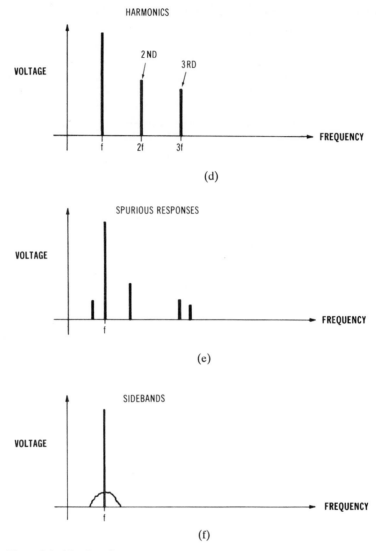

Figure 3-4 *(Continued)*

3.3.4 Distortion

The waveform may have zero frequency error and zero amplitude error, but have considerable DISTORTION. Distortion is any imperfection in the shape of the waveform as compared to the desired signal. In the sine wave case, distortion shows up in the frequency domain as undesirable harmonics, which is referred to as HARMONIC DISTORTION. Harmonic distortion is shown in Figure 3-4d with the original pure sine wave now accompanied by several frequency components at integer multiples of the original frequency. The maximum amplitude of the harmonics may be specified as a percentage of the fundamental or in dB relative to the fundamental.

All the harmonics may be lumped into one distortion number called TOTAL HAR-MONIC DISTORTION (THD) and specified in percent.

3.3.5 Spurious Responses

A signal source may produce low-level frequency components that are not harmon-ically related to the output frequency. These components are called SPURIOUS RESPONSES and are often the result of the particular signal-generation technique used inside the instrument. Alternatively, they may result from the AC power line frequency appearing either directly or as modulation on the carrier. Spurious re-sponses may appear at any frequency and may move when the source's frequency changes. Figure 3-4e shows the effect of spurious responses in the frequency domain. The maximum value of this type of response is usually specified at an absolute level or in dB relative to the desired signal (or both).

3.3.6 Close in Sidebands

If the sine wave were absolutely pure, in the frequency domain it would have an infinitely thin spectral line, indicating that all the signal's energy is at exactly one frequency. In reality, the sine wave's frequency response may spread out slightly, often causing a pedestal effect as shown in Figure 3-4f. These imperfections usually appear as noise sidebands very close to the fundamental and are thought of and specified in a variety of ways. Some instruments specify these sidebands as RESIDUAL MOD-ULATION (AM or FM), while others refer to the phenomenon as PHASE NOISE. These different mechanisms are not exactly equivalent, but they all produce a noiselike response close to the output frequency.

3.4 SINE WAVE SOURCES

As the name implies, sine wave sources are capable of supplying only sine wave signals. These sources are economical instruments that generate low distortion signals ranging in frequency from a few Hz to about 1 MHz. Internally, the technology used is usually a free-running oscillator operating at the same frequency as the desired output frequency. This technology is inexpensive, but is limited in frequency range. Sine wave oscillators are used mainly for audio frequency work, with a 600-ohm output impedance being very common. Some sine wave sources are designed to have very low distortion, which is important when measuring distortion in audio amplifiers.

Figure 3-5 shows a simplified block diagram of a sine wave source. It consists of a free-running oscillator followed by an amplifier to boost the signal level and an attenuator to vary the output level. The free-running oscillator frequency is usually controlled by a variable capacitor and/or resistor. Figure 3-6 is an example of sine wave source, this one having an auxiliary square wave function.

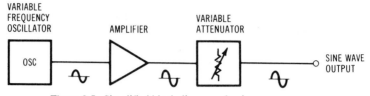

VARIABLE
FREQUENCY
OSCILLATOR   AMPLIFIER

VARIABLE
ATTENUATOR

OSC

SINE WAVE
OUTPUT

Figure 3-5 Simplified block diagram of a sine wave source.

Figure 3-6 A sine wave source that also has a square wave output. (Photo courtesy of Leader Instruments Corporation.)

3.5 FUNCTION GENERATORS

Function generators are the most widely used general-purpose signal source. They are capable of supplying nonsinusoidal waveforms such as square waves, triangle waves, and pulse trains in addition to sine waves (Figure 3-7). Depending on the particular instrument, a function generator may also provide modulation and swept-frequency capability. The function generator often uses technology similar to the sine wave oscillator. A simplified block diagram is shown in Figure 3-8. One of the required waveforms is produced by a free-running oscillator and then conversion circuits derive the other waveforms from the original. In Figure 3-8, a triangle wave is generated by the oscillator. A square wave is derived from the triangle wave by running the triangular wave through a comparator. (The comparator may actually be part of the triangle wave oscillator.) A sine wave is derived from the triangle wave by passing it through a wave-shaping circuit (sine shaper). The desired waveform is selected, amplified, and output through a variable attenuator. As the technology has improved, the function generator has gradually replaced the sine-wave-only source since it offers sine wave capability along with the other waveforms to provide much greater flexibility.

Function generators typically provide a DC offset adjustment that allows the user to add a positive or negative DC level to the generator output. Figure 3-9 shows how adding varying amounts of direct current to a square wave can produce different

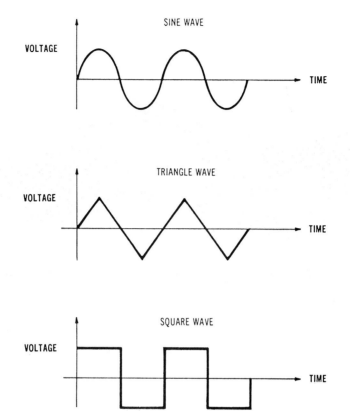

Figure 3-7   Standard function generator waveforms.

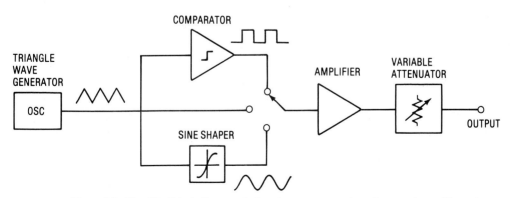

Figure 3-8   Simplified block diagram of a function generator using a free-running oscillator that generates a triangle wave.

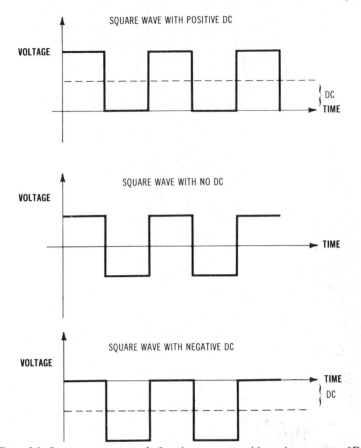

Figure 3-9   Square wave output of a function generator with varying amounts of DC offset added.

waveforms. The DC offset can also be used to control the DC bias level into a solid-state circuit.

Function generators can be used for audio sine wave testing just as the sine-wave-only generator can. In addition, the square wave output can be used as a clock for digital circuits. Some function generators include a fixed TTL compatible output for such an application. On others, square wave output and DC offset adjustment can be used to generate valid logic levels. The triangle wave can be used in situations where a ramplike voltage is needed. Figure 3-10 shows an example of a typical function generator.

3.6 PULSE GENERATORS

Pulse generators are specifically designed to produce high-quality square waves and pulse trains. They generally operate over a frequency range as low as 1 Hz and as high as 1 GHz. As shown in Chapter 1, a pulse train has a large number of significant

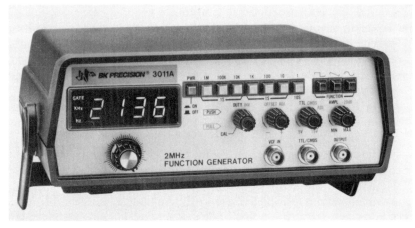

Figure 3-10   A typical function generator. Photo courtesy of B & K Precision.

harmonics requiring an instrument bandwidth much greater than the fundamental frequency. General-purpose pulse generators which produce frequencies up to about 50 MHz can often generate function generator waveforms, too. Above 50 MHz, pulse generators are usually optimized for the difficult task of producing clean pulses. The bandwidth required to generate these pulses extends to hundreds of megahertz, so high-performance pulse generators tend to have a correspondingly high price. Figure 3-11 shows a 50-MHz pulse generator.

Figure 3-11   A pulse generator capable of producing frequencies up to 50 MHz. Photo courtesy of Hewlett-Packard Company.

For convenience, pulse generators are generally specified in the time domain. Waveform characteristics such as period, duty cycle, or pulse width can be selected from the front panel. Distortion is not specified since it is basically a sine wave type of specification. Instead, pulse-related parameters like rise time and fall time are specified. Usually very flexible control is included to allow positive-going pulses, negative-going pulses, and symmetrical waveforms (Figure 3-12).

Pulse generators may supply a trigger circuit which produces an output pulse in response to a trigger signal. Figure 3-13 shows how a trigger signal initiates an output pulse with a controllable amount of delay in between. This feature can be used to simulate special conditions in digital logic circuits. The trigger condition may also come from the push of a front panel button. This is handy for single stepping a sequential logic circuit (a microprocessor, for example). Some pulse generators provide burst modes which allow a fixed number of pulses to be output repetitively or in response to a trigger. Triggering capability generally results in lower frequency stability for pulse generators (as compared with other signal sources).

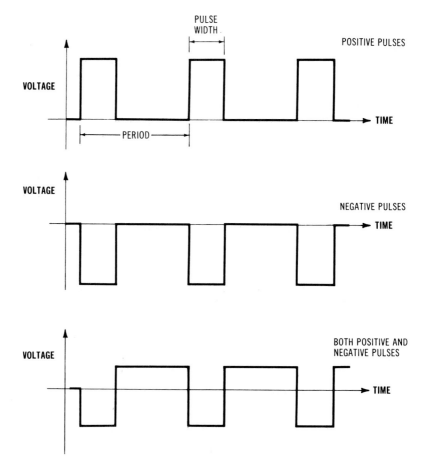

Figure 3-12 Flexible control over the polarity of the output pulses is provided on a pulse generator.

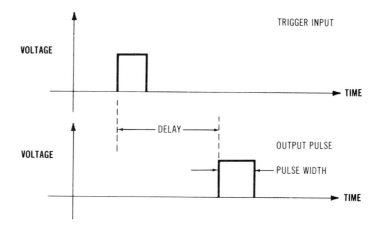

Figure 3-13 The trigger input causes a pulse at the generator's output after a controllable amount of delay.

Some pulse generators provide a fixed TTL-compatible logic signal for use with digital circuits. This is, of course, not absolutely necessary since any reasonable pulse generator can be configured to put out logic levels, but it is convenient to have a fixed logic output predefined.

Pulse generators are generally used in digital systems to simulate or replace digital signals such as clocks and data and control lines. The burst and triggering features allow the user to clock a digital system a known number of times. This is very handy for single stepping a piece of digital hardware when troubleshooting. The pulse generator can also be used in analog systems. Square waves are used to test audio amplifiers and related equipment (see Chapter 5).

3.7 SIGNAL GENERATORS

A signal generator is primarily designed to be used as a test stimulus for radio receivers, but also ends up being used as a general-purpose sine wave source in the radio frequency range. Signal generators are provided with flexible modulation capability, often with a built-in audio modulating source. Amplitude modulation and frequency modulation are most common, with phase and pulse modulation also sometimes provided (Figure 3-14). Precise, wide-range attenuation and low-RF leakage are usually required in signal generators to allow full-range receiver tests. Figure 3-15 shows a typical signal generator.

Figure 3-16 is the conceptual block diagram of a signal generator having an internal audio oscillator for use as a source of modulation. The heart of the signal generator is a voltage controlled oscillator (VCO). The frequency of the VCO is determined by the voltage present at the control input. If the control voltage is increased or decreased, then the frequency of the VCO increases or decreases, respectively. So whatever signal is applied to the control voltage shows up as the frequency of the oscillator. This is exactly what is required for frequency modulation. The audio

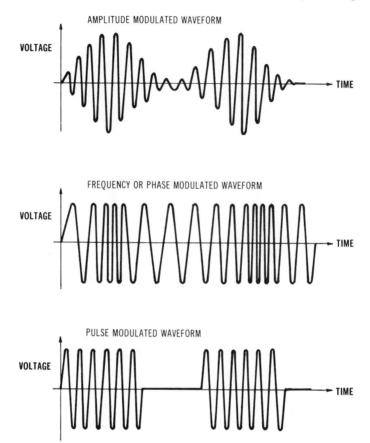

Figure 3-14 Typical waveforms produced by signal generators.

Figure 3-15 A radio frequency signal generator with an internal modulation source.
Photo courtesy of Hewlett-Packard Company.

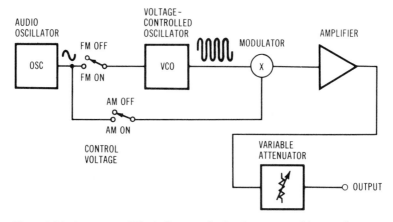

Figure 3-16 A conceptual block diagram of a signal generator with an audio source for modulating the carrier.

modulating signal drives the control input of the VCO to produce a frequency modulated carrier. For amplitude modulation, a modulator circuit is added after the oscillator. The modulator circuit varies the amplitude of the VCO's output without modifying its frequency. Thus, the output is an amplitude modulated signal. The actual block diagram used to implement the signal generator may be much more complicated, but the conceptual block diagram is still a valid tool for these signal sources.

The frequency accuracy and stability are very important in a signal generator used to test receivers. The passband of the receiver is usually very narrow, much less than a percent of the center frequency. An even better frequency accuracy is required for the test signal. Harmonic distortion is not usually very critical in receiver testing since the harmonics are well away from the frequency of the receiver. (For other high-frequency applications, however, the harmonic distortion may be important.) Amplitude accuracy is important for accurately measuring the receiver's sensitivity.

Higher-quality signal generators have extremely good close-in sideband performance (usually specified as residual FM or phase noise). This is important in radio and other systems where even very low levels of noise close to the carrier can degrade the system performance. For instance, the rejection of a signal that is very close to but outside a radio receiver's bandwidth is an important test of a receiver's performance. This adjacent channel rejection test can be performed by tuning the signal generator to a frequency very close to the receiver's frequency and then increasing the generator's output level until the receiver can no longer reject the signal. The signal generator must not have any close-in sidebands; otherwise, these sidebands will spill over into the receiver passband and be detected as if they were a valid signal.

3.8 SWEEP GENERATORS

Sweep generators are sine wave sources that have the ability to change their frequency in a controlled manner. This swept-frequency capability is useful for testing circuits over a wide frequency range in a short amount of time. Of course, the sweep feature can be disabled and the source can be used to produce fixed-frequency sine waves. The sweep generator is often combined with a function generator to provide both types of operation in one instrument. Some function generators go so far as to supply sweep, pulse, and modulation capability in one box. A function generator that also includes sweep capability is shown in Figure 3-17.

Sweep generators generally sweep in frequency in a linear manner but may also be able to sweep logarithmically. A sweep voltage output which is proportional to the frequency during a sweep is usually provided. This output can be used to drive other instruments, particularly an oscilloscope configured for a frequency response test (see Chapter 5). (The sweep voltage is often called the X-drive output, since it is used to drive the X-axis of the oscilloscope display.) Figure 3-18 shows a sine wave being swept in frequency along with its corresponding sweep voltage. Figure 3-19 shows a simplified block diagram of a sweep generator. A voltage controlled oscillator is driven by the sweep voltage to produce a frequency sweep. The output of the VCO is amplified and passes through a variable attenuator.

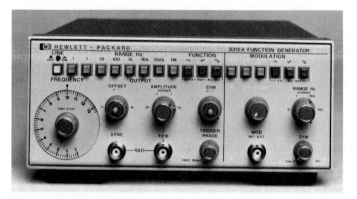

Figure 3-17 A function generator with sweep and modulation capability. Photo courtesy of Hewlett-Packard Company.

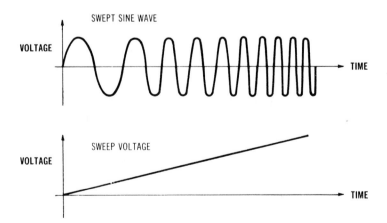

Figure 3-18 The output of a sweep generator is a sine wave that increases in frequency. The sweep voltage is proportional to the output frequency.

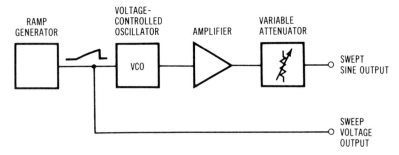

Figure 3-19 Simplified block diagram of a sweep generator.

3.9 ARBITRARY WAVEFORM GENERATORS

A special class of sources has recently emerged, which have the ability to generate arbitrary waveforms. A simplified block diagram is shown in Figure 3-20. Digital data representing the desired waveform is stored in digital memory (waveform memory). An address counter sequences through the addresses of the memory, causing the data in the memory to appear sequentially at the digital port of a digital-to-analog converter (DAC). The DAC produces a voltage proportional to the digital data supplied to it. Thus, as the address cycles through memory, the waveform data is transformed into a voltage waveform at the output of the DAC. A VCO (or other oscillator) is used to clock the address counter at a variable rate, to allow the waveform period (frequency) to be varied. The faster the address counter is clocked, the faster the memory is cycled through and the higher the waveform frequency. The output of the DAC is amplified and then passed through a variable attenuator to the output.

An arbitrary waveform generator can be used in a variety of applications where previously it was difficult or impossible to generate appropriate waveforms. Of course, it can also be used to generate any of the waveforms previously discussed (square,

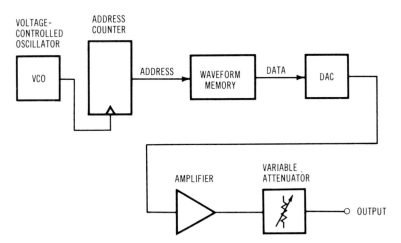

Figure 3-20 The simplified block diagram of an arbitrary waveform generator.

triangle, pulse), but the real utility of the generator is in simulating more complex signals. Figure 3-21 shows some examples of waveforms that can be produced using an arbitrary waveform generator. Imperfections in signals (such as overshoot in a square wave and glitches in a digital signal) can be simulated in a controlled manner to determine how sensitive a circuit is to such problems. Also, transient signals such as a damped sine wave can be produced. Another advantage of the arbitrary waveform generator is the ability to change frequency quickly and precisely (frequency hop).

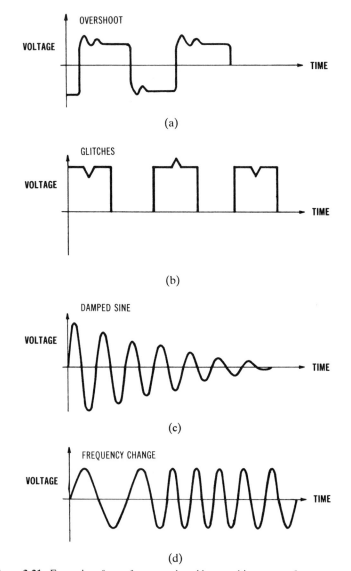

Figure 3-21   Examples of waveforms produced by an arbitrary waveform generator. (a) Square wave with overshoot. (b) Digital signals with glitches. (c) Damped (decaying) sine wave. (d) Sine wave with abrupt frequency change.

The application of arbitrary waveform generators is limited only by the number of different signals present in the world. Any waveform that can be stored in digital form and loaded into the waveform memory can be simulated. Of course, the arbitrary generator is still limited by the same bandwidth constraints that limit other instruments. Also, the DAC performance, the address counter clock rate, and the size of the waveform memory will limit the frequency range of the generator. The actual implementation of the arbitrary waveform generator varies from compact, stand-alone instruments to generators that are dependent on an external computer to load the waveform data. Many generators supply a "waveform calculator" which allows the user to define the arbitrary waveform easily, without manually entering each waveform point.

3.10 SYNTHESIZED SOURCES

In a traditional signal source, the waveform is produced by one or more free-running oscillators which are adjusted to operate over some frequency range by varying the value of a component, usually combinations of capacitors, inductors, and resistors. In a synthesized signal source, one or more crystal-controlled reference oscillators operate at a fixed frequency which is then used by the synthesizer circuitry to produce the desired frequency (Figure 3-22). Because the reference oscillator does not have to change frequency, it can be made extremely stable. The synthesizer consists of digital dividers and other circuits which have no frequency variation associated with them. So the source frequency is as stable and accurate as the reference oscillator. The result is a variable frequency source that has the precise frequency control that is normally only achievable with a fixed-frequency oscillator.

Synthesized sources do not usually have a single-frequency accuracy specification. Instead, the reference oscillator's frequency drift is specified (usually in parts per million per month or year). The discussion in Chapter 6 of frequency counters explains this type of specification.

3.10.1 Frequency Locking Multiple Sources

The reference oscillator is usually designed such that it can be electronically locked to other synthesized sources. Each source provides a reference output (usually 1, 5, or 10 MHz) as well as a reference input. The reference output of one source is connected to the reference input of one or more other sources (Figure 3-23). These

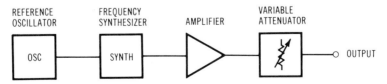

Figure 3-22 Simplified block diagram of a synthesized source has a reference oscillator that is used by the synthesizer to produce the desired output frequency.

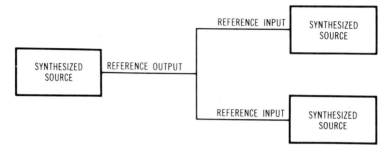

Figure 3-23. Multiple synthesized sources can be locked together in frequency using the reference inputs and reference outputs.

sources lock their internal reference oscillators to the reference output of the original source. This results in all the sources being locked precisely in frequency to one reference oscillator. Two synthesized sources that are locked together exhibit essentially zero frequency drift between them. (Two free-running sources will drift apart in frequency quite noticeably.)

Some sources are simply called synthesizers because their primary purpose is precise frequency synthesis. However, as technology has improved, frequency synthesis techniques have been used to implement most types of signal sources, especially signal generators, sweep generators, and function generators. These instruments perform the same basic function as previously outlined, but when synthesized, their frequency accuracy and stability are greatly improved. Figure 3-24 shows a synthesized source that combines function generator waveforms, frequency sweep capability, and modulation into one instrument with synthesizer frequency accuracy.

Table 3-1 summarizes the basic characteristics of the various types of sources. Because of the overlap often found between the categories of sources, this table is only a general guide as to the specifications and capabilities of each type of source.

Figure 3-24 A synthesized source with function generator, sweep and modulation capability, Photo courtesy of Hewlett-Packard Company.

TABLE 3-1 TABLE OF SIGNAL GENERATOR CHARACTERISTICS. THIS TABLE REFLECTS THE
SPECIFICATIONS OF A TYPICAL INSTRUMENT IN EACH CATEGORY. ACTUAL SPECIFICATIONS
AND CAPABILITY WILL VARY FROM INSTRUMENT TO INSTRUMENT.

Generator	Freq. Range	Freq. Accuracy	Harmonic Distortion	Sweep	Modulation	Sine	Square/ Pulse
Sine wave	1 Hz– 1MHz	5%	0.1%	No	No	Yes	No
Function generator	1 Hz– 10 MHz	5%	1%	Some models	Some models	Yes	Yes
Pulse generator	1 Hz– 1GHz	5%	Not applicable	No	No	No	Yes
Signal generators	100 kHz –1GHz	0.005%	2%	No	Yes	Yes	No
Sweep generators	10 Hz– 30 MHz	5%	1%	Yes	No	Yes	Some models

Example 3-2.

Of the types of sources described in this chapter, which ones would suit the following applications?

1. Generate a 1-MHz digital clock.

2. Provide a 2-KHz sine wave for audio testing.

3. Generate a 29.6-MHz frequency modulated sine wave for testing a radio receiver.

1. Any source capable of outfitting a square wave, with perhaps some DC offset: function generator or pulse generator.

2. All the sources except the pulse generator can output a sine wave, but signal generators usually don't go that low in frequency: sine wave source, function generator, sweep generator.

3. Although modulation is sometimes supplied in other sources, they typically don't go high enough in frequency: signal generator.

REFERENCES

HELFRICK, ALBERT D. and WILLIAM D. COOPER. *Modern Electronic Instrumentation and Measurement Technique*, Englewood Cliffs, N.J.: Prentice Hall Inc., 1990.

JONES, LARRY D. and A. FOSTER CHIN. *Electronic Instruments and Measurements*, 2nd Edition, Englewood Cliffs, NJ: Prentice Hall Inc., 1991.

OLIVER, BERNARD M., and JOHN M. CAGE. *Electronic Measurements and Instrumentation*. New York: McGraw-Hill Book Company, 1971.

"Signal Generator Spectral Purity," Hewlett Packard Company, Application Note 388, Publication Number 5952-2019, 1990.

chapter four

Oscilloscopes

The oscilloscope is undoubtedly the most versatile electronic instrument in use today. Unlike meters, which only allow the user to measure amplitude information, the oscilloscope (or "scope") allows the user to view the instantaneous voltage versus time. This reveals the shape of the waveform and parameters such as frequency and phase can also be measured.

4.1 THE OSCILLOSCOPE CONCEPT

The primary function of an oscilloscope is to display an exact replica of a voltage waveform as a function of time. This picture of the waveform can be used to determine quantitative information such as the amplitude and frequency of the waveform as well as qualitative information such as the waveform's shape. The oscilloscope can also be used to compare two different waveforms and measure their time and phase relationships.

The initial discussion of oscilloscopes will be centered on the typical two-channel analog scope (Figure 4-1). Single-channel scopes used to be available, but have largely been eclipsed by the two-channel variety. Four-channel scopes are also prevalent in the marketplace and are particularly useful for measuring signals in digital systems. (Digital systems usually have multiple data and address lines that may need to be viewed simultaneously.) Four-channel scopes operate similarly to two-channel scopes, with the exception of having twice as many inputs and vertical amplifiers. For general-purpose use below 100 MHz, analog scopes are usually the most economical alternative, but digital oscilloscopes have become as affordable as their analog counterparts. In concept, analog oscilloscopes and digital oscilloscopes perform the

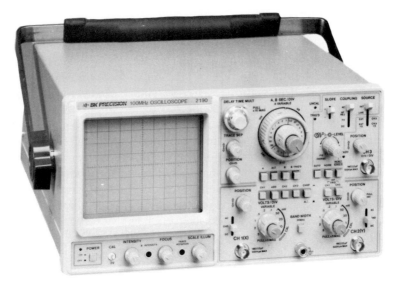

Figure 4-1 A typical two-channel 100-MHz analog oscilloscope. (Photo courtesy of B & K Precision—Dynascan corporation.)

same types of measurements, but with different techniques internal to the instrument. These differences will be addressed later in the chapter.

A sine wave is displayed by an oscilloscope as shown in Figure 4-2. The vertical axis represents voltage and the horizontal axis represents time. This results in the same voltage versus time plot for a sine wave that was introduced in Chapter 1. The oscilloscope display has a set of horizontal and vertical lines called the GRATICULE which aids the user in estimating the value of the waveform at any particular point. (The graticule is shown in Figure 4-2, but is omitted in other figures.) Fundamentally, the concept of an oscilloscope is simple: Accurately produce the voltage versus time plot of a waveform. Actually accomplishing this may be more complicated.

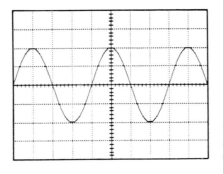

Figure 4-2 The oscilloscope display with a sine wave connected to the oscilloscope input.

4.2 TIMEBASE OPERATION

A simplified block diagram of an oscilloscope is shown in Figure 4-3. The display of the oscilloscope (usually a CATHODE RAY TUBE or CRT) requires two pieces of information: the vertical position to be plotted and the horizontal position to be plotted. The vertical axis represents the waveform to be plotted, so the input of the oscilloscope is connected to an amplifier which drives the vertical position of the display. Time is plotted on the horizontal axis. Driving the horizontal position of the display is an internally generated waveform that represents time. The section of circuitry that produces the time waveform is called the TIMEBASE. We know what the vertical signal looks like—it's just whatever the input waveform is, perhaps a sine wave or square wave. What we want the timebase to do is constantly increase the horizontal position as the input voltage goes through a cycle. This results in the input waveform being swept or painted across the display at some rate, depending on the timebase. The timebase waveform required to do this is a constantly increasing ramp voltage. Since the display is usually updated repetitively, the ramp starts over when it reaches its maximum value, resulting in a sawtooth waveform (Figure 4-4). Each cycle of the sawtooth waveform corresponds to one sweep across the display. A small delay (called the RETRACE TIME) occurs between each ramp of the sawtooth as the horizontal position is reset to the left side of the display.

4.3 TRIGGERING

A timebase alone is not sufficient to produce a usable display. The timebase needs to know when to start a sweep. Without this information, the display will attempt to show the correct waveform, but with a random start time. At best, this results in a unstable waveform wandering across the display. At worst, the display is a jumbled collection of waveforms smeared together filling the entire display area (Figure 4-5a).

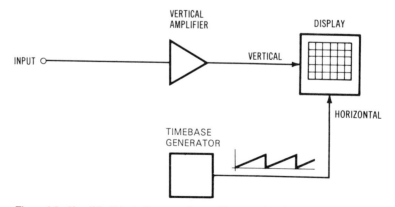

Figure 4-3 Simplified block diagram of an oscilloscope showing timebase operation.

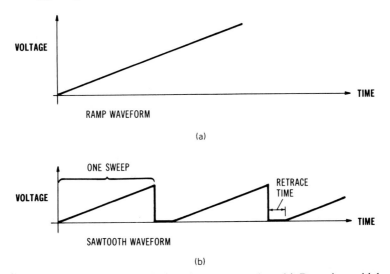

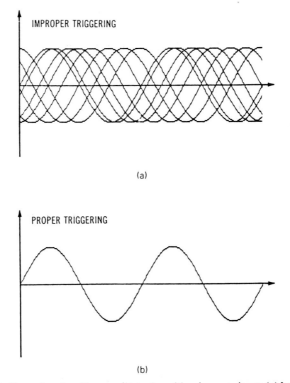

Figure 4-4   The timebase generator produces a ramp voltage (a). Repeating multiple ramps results in a sawtooth waveform (b).

Figure 4-5   Examples of oscilloscope triggering with a sine wave input. (a) Improperly triggered waveform results in an unstable display. (b) Properly triggered waveform has a stable display.

A properly triggered waveform, having a predictable and repeatable starting point, will be stationary from sweep to sweep (Figure 4-5b). Consider the sine wave shown in Figure 4-6. If the trigger circuit is set up to trigger at the start of the sine wave, then triggers will occur at the points identified in the figure. Each piece of the sine wave is then displayed, with the trigger point at the leftmost edge of the display. Notice that each trigger point occurs at the same place on each cycle. This will result in a stable display as shown in Figure 4-6b. The figure implies that none of the cycles of the waveform is missed by the oscilloscope. In reality, since the oscilloscope may not be able to prepare for a new sweep instantaneously (due to retrace time), some cycles may be lost. With waveforms that are relatively consistent, this is not a problem.

Triggering is probably the most troublesome part of using an oscilloscope. Most scopes provide a variety of ways to trigger on a signal so that the user can customize the triggering to a particular measurement problem. Of course, this also means that the user must understand and make decisions about what type of triggering to use.

The trigger point on a waveform is usually defined using both a voltage level and a slope (positive or negative). The TRIGGER LEVEL control determines the voltage level at which the trigger will occur. The TRIGGER SLOPE control determines whether the trigger will occur on rising (positive slope) or falling (negative slope) portions of a waveform. Figure 4-7 shows a waveform and some possible trigger times. All the trigger points shown are at a trigger level of 0.5 volts, but some are for a positive slope, and the others are for a negative slope. The resulting scope displays for the two triggering conditions are shown in Figures 4-7b and 4-7c.

When the waveform being measured is used as the trigger source, the oscilloscope is using INTERNAL TRIGGER. Multiple-channel scopes allow the user to select which of the channels is to be used as the trigger source. Internal

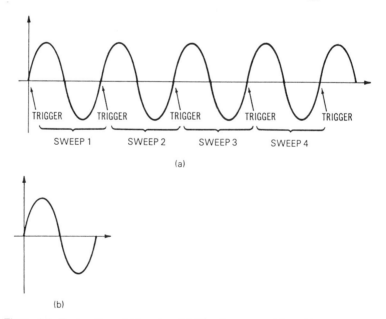

Figure 4-6   Explanation of triggering. (a) The sine wave with multiple cycles and multiple trigger points. (b) The resulting properly triggered oscilloscope display.

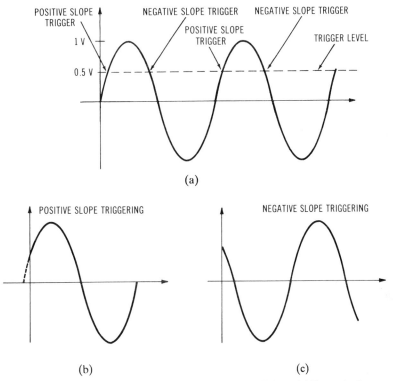

Figure 4-7 Oscilloscope trigger control includes level and slope. (a) Example show-ing both positive slope and negative slope trigger points. (b) The scope display if positive slope triggering is used. (c) The scope display if negative slope triggering is used.

trigger is used the most, but it is possible to trigger on other waveforms. An EXTERNAL TRIGGER input is provided to allow the user to connect an external signal to the oscilloscope to be used for triggering purposes. This signal usually cannot be viewed on the scope display, but some scopes have this capability to help the user in setting up the triggering. The LINE TRIGGER selection uses the AC power line voltage as the trigger signal (usually 60 Hz). Line trigger is useful for observing waveforms that are either directly related to the power line frequency (including its harmonics) or have power line–related voltages superim-posed on the original waveform.

The trigger may be AC coupled or DC coupled. DC coupling presents the trigger circuit with a waveform containing both AC and DC voltages, while AC coupling removes any DC that may be present. AC coupling is useful when the desired trigger signal is riding on top of a DC voltage which needs to be ignored. Many oscilloscopes include additional filters that can be switched into the trigger circuit to condition the trigger signal. The LOW-FREQUENCY REJECT filter will remove low frequencies such as the 60-Hz line frequency (and its harmonics) which may be present in the trigger signal, causing triggering problems. The HIGH-FREQUENCY REJECT filter will similarly remove high-frequency noise which also may cause triggering problems.

4.3.1 Trigger Holdoff

The TRIGGER HOLDOFF control disables the triggering circuit for a period of time after the end of a sweep. This is useful when the waveform has several places in its cycle that are the same as the trigger condition. The oscilloscope would normally trigger on all of them, but with the trigger holdoff, the scope ignores all but the first one. Suppose we want to display the first cycle of the pulsed sine wave shown in Figure 4-8. The trigger could be set to zero volts and positive slope, which will trigger at the beginning of the sine wave. Unfortunately, this trigger condition exists at the start of every one of the sine wave periods, so the oscilloscope will display every cycle. However, if the trigger holdoff is adjusted properly, the subsequent triggers can be ignored and only the first cycle of each sine pulse will be displayed.

While setting the trigger holdoff as a fixed time (TIME HOLDOFF) is most common, some scopes provide an additional holdoff feature, called EVENT HOLD-OFF. Instead of specifying the amount of holdoff as a fixed time, the user specifies the number of trigger events that need to be ignored. This is useful in cases where a fixed number of pulses needs to be ignored and the total time associated with these pulses varies significantly (such as pulses from a disk drive that is varying in speed).

4.3.2 Video Trigger

Since composite television and video signals can be quite complex, they don't lend themselves to simple edge triggering. Therefore, many scopes provide a TV or video trigger capability which extracts the horizontal and vertical sync signals from the input video and uses them for triggering the scope. The simplest form of video triggering lets the user trigger on all horizontal sync pulses (all lines) or all vertical sync pulses (all fields). More advanced triggering systems let the user pick the individual line and field to trigger on, allowing any particular portion of the video waveform to be easily displayed.

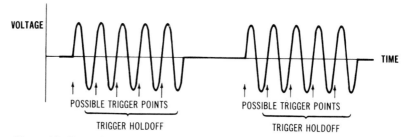

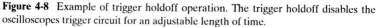

Figure 4-8 Example of trigger holdoff operation. The trigger holdoff disables the oscilloscopes trigger circuit for an adjustable length of time.

4.3.3 Logic Trigger

Edge triggering is sufficient for many simple signals, but the complex signals found in high-speed digital and microprocessor systems require a more sophisticated triggering system. Logic triggering expands on the usual level and slope triggering by letting the user define the trigger condition in more powerful terms. Each of the scope channels is considered as a logic level, with a logic high (H) being any voltage greater than the trigger level and a logic low (L) being any voltage below the trigger level. These trigger levels can be set independently on each channel, to accommodate a mixture of logic levels. Any scope input can be ignored for triggering purposes by specifying its trigger condition as a "don't care" (X).

Although logic triggering can be implemented on either an analog or digital scope, it tends to be found only on digital scopes. This is because the pretrigger viewing capability of the digital scope works well with logic triggering to measure and test digital circuits.

PATTERN TRIGGER lets the user define a trigger condition by specifying a logic pattern which is to be present on the scope's inputs. As soon as this pattern appears, the scope triggers. Figure 4-9 shows the input waveforms on the four input channels of a scope. With the trigger level set near the middle of the logic swing, the scope is set to pattern trigger on the pattern H H L L (high, high, low, low) on channels 1 through 4. The scope triggers at the point shown in Figure 4-9.

This feature can be used to trigger on such things as a particular digital value on an address or data bus. Suppose that in a microprocessor system, writing to a peripheral device is causing a system failure. Pattern trigger could be used to trigger on the appropriate address and enable lines associated with writing to the device. Once the trigger condition is established, the scope can be used to view other signals in the system.

A more powerful version of pattern triggering is TIME-QUALIFIED PATTERN TRIGGERING. In addition to defining a logic pattern, the user specifies how long the pattern must be present for a valid trigger to occur. The time limit can be specified as greater than a limit, less than a limit or within some range. A simple application of this feature is triggering on a glitch (Figure 4-10). Suppose that the normally low output of a digital gate produces a high-going glitch which is suspected of causing a system failure. The scope is set to pattern trigger on H X X X (high, don't

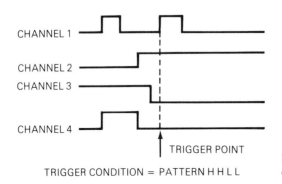

TRIGGER CONDITION = PATTERN H H L L

Figure 4-9 Example of pattern trigger operation.

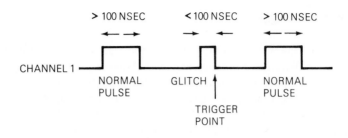

TRIGGER CONDITION = PATTERN H X X X PRESENT < 100 NSEC

Figure 4-10 Time-qualified pattern trigger can be used to trigger on glitches.

care, don't care, don't care) with a time qualifier of <100 nsec. Channel 1 is connected to the digital signal, and when a glitch occurs with duration less than 100 nsec, the scope triggers. This assumes that legitimate signals pulse high for more than 100 nsec. If the legitimate pulses are of a different duration, the time qualifier can be increased or decreased so that legitimate pulses are ignored and only glitches are triggered on. Of course, the other channels which are set to "don't care" in our example can be used to qualify further the glitch. For instance, we may be interested only in glitches that occur when a particular enable or read/write line is low or when a particular pattern appears on a control bus.

STATE TRIGGER configures the trigger system somewhat like a logic analyzer in that one channel of the scope is defined as the clock of the trigger system and the other channels are read on the rising or falling edge of the clock. For example, channel 1 might be configured as the clock (rising edge) and the trigger condition on the other channels might be H H L (Figure 4-11). On each rising edge on channel 1, the trigger

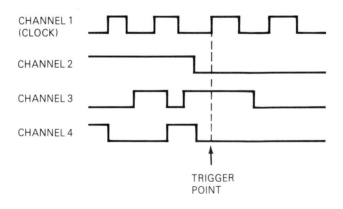

TRIGGER CONDITION = STATE ↑ L H L

Figure 4-11 State trigger triggers on a pattern, but only on clock edges.

system evaluates the logic level of each channel, and when the state of the channels matches the trigger condition, a valid trigger has occurred.

State triggering has the most utility in situations where logic levels are only valid on a clock edge. For example, a D-type flip-flop changes its output based on the state of its D input when a rising edge appears on its clock line. The state of the D input is significant only when a rising clock edge occurs. Time-qualified pattern trigger could be used to trigger on some event such as a glitch at the D input, but it would not be useful for triggering on a particular logic state *valid at the clock edge*. However, state trigger fires *only* at the clock edges and *only* when the specified logic state occurs simultaneously.

4.4 SWEEP CONTROL

In the SINGLE-SWEEP mode, the oscilloscope displays only one sweep in response to a trigger and then waits until the user resets the scope for another sweep. This allows the user to capture a particular single-shot, or transient event, rather than continuously updating the display with unwanted information. Although this feature is supplied on almost all oscilloscopes, it is most useful when the scope also provides a method of storing the waveform, at least temporarily, on the screen. Otherwise, the single-sweep mode tends to be a brilliant, but brief, flash across the display.

4.4.1 Automatic and Normal Sweep

Most oscilloscopes supply two types of continuous sweeps: AUTOMATIC and NORMAL. The difference between the two is rather subtle, but important. With normal sweep, the oscilloscope is swept when a valid trigger occurs. This is fine for most AC waveforms, but it is inconvenient if the input is a DC voltage. In this case, no trigger will occur, since the DC input is constant and will not be passing through whatever trigger level happens to be selected. The result is a blank display. Automatic sweep operates just like normal sweep except that if no trigger occurs for a period of time (as in the case of a DC input or an incorrectly adjusted trigger level), the scope generates a sweep anyway. This is convenient because it gives the user a look at the waveform, even if it isn't triggered properly. The user can then determine whether the waveform is on screen and what action is necessary for correct triggering. In the case of a pure DC voltage, triggering is rather meaningless—the user just wants to view the constant voltage. If a trigger starts occurring at a fast enough rate (typically anything greater than 40 Hz), then the scope triggers just like the normal sweep mode. Automatic sweep should be the default choice since it will give good results on most waveforms. The exception is when the desired trigger condition occurs so infrequently (a very-low-frequency sine wave, for example) that the automatic mode will begin triggering before the appropriate time. In that case, normal sweep should be used.

4.4.2 Chop and Alternative Modes

Most oscilloscopes have at least two channels, both of which can be displayed at the same time. One way to do this is to have a display that can plot two waveforms simultaneously. Analog oscilloscopes that have true simultaneous dual-trace capability are called DUAL-BEAM OSCILLOSCOPES. This is because they use a CRT display that has two electron beams active at the same time. Another, more cost effective, method is to use a single-beam display but switch electronically between the two input signals (Figure 4-12). If this is done quickly enough, it will not be noticeable to the user, nor will it affect most measurements. (This consideration does not apply to digital scopes which are not limited by the type of display, but by the analog-to-digital conversion circuitry.)

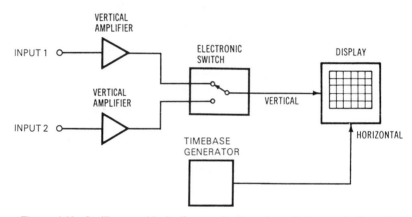

Figure 4-12 Oscilloscope block diagram for two-channel chop and alternate modes. The electronic switch is used to display both channels (almost) simultaneously.

There are two different methods of controlling the electronic switch. CHOP mode switches between the two inputs as fast as possible while the waveforms are being plotted on the display. Figure 4-13 shows how chop mode can be used to display two different waveforms simultaneously. For clarity, channel 1 is shown as a triangle wave, and channel 2 is shown as a square wave. The display trace switches back and forth ("chops") between the two waveforms during the sweep, resulting in both of them appearing on the screen. Figure 4-13 exaggerates the chop effect to illustrate the point. In practice, the two channels are switched fast enough so that the chopping cannot be seen. This works well for sweeps that are much slower than the rate at which the two channels chop, so chop mode is best for low-frequency signals. If chop mode is used on high-frequency waveforms (and with very fast

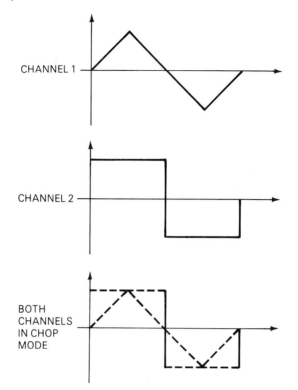

Figure 4-13 In chop mode, the display switches between the two channels as fast as possible, to display both waveforms simultaneously without any noticeable chopping.

sweeps), then the chopping effect is noticeable on the display (as illustrated in Figure 4-13).

On the other hand, ALTERNATE mode takes a complete sweep of one waveform, then switches to the other input and takes another sweep. As shown in Figure 4-14, the first sweep displays channel 1, the second sweep displays channel 2, the third sweep displays channel 1 again, and so on. This works well during fast sweeps, because if the update rate to the display is fast, the user perceives that both inputs are being plotted simultaneously. Therefore, alternate mode works best with high-frequency signals. If alternate mode is used on low-frequency waveforms (and with slow sweeps), the individual sweeps of each channel become apparent, and the effect of simultaneously displayed channels is lost. One potentially misleading problem with alternate mode is that even though the two waveforms appear to be displayed simultaneously, they really were measured at two distinct points in time (on two different triggers). For most measurements, this is not a problem since the waveform repeats the same cycle every time. There are cases, however, where the user needs to measure both channels truly simultaneously.

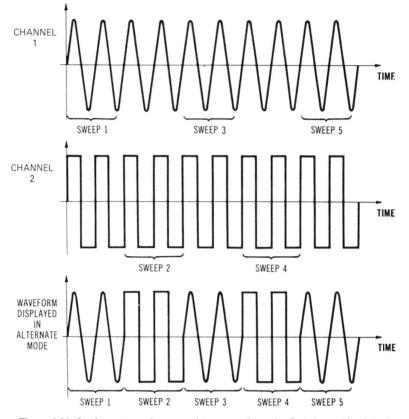

Figure 4-14 In alternate mode, a complete sweep from the first channel is plotted, and then a complete sweep from the second channel is plotted. For fast sweeps, this is done quickly enough to create the effect of two simultaneous waveforms.

4.5 VERTICAL AMPLIFIER

The gain of the vertical amplifier shown in Figure 4-12 determines how big the waveform appears on the display. The VERTICAL SENSITIVITY control determines the gain of the vertical amplifier and is calibrated in volts per division so that the user can determine the amplitude of the signal by counting the number of vertical divisions on the display. For example, the waveform shown in Figure 4-2 showed a sine wave that is four vertical divisions peak-to-peak. If the vertical sensitivity control was set to 0.2 volts per division, then the sine wave would be $4 \times 0.2 = 0.8$ volts peak-to-peak.

Oscilloscopes have a vertical amplifier for each of their input channels. In the usual timebase mode of the oscilloscope, both channels are displayed as a function of time. As stated previously, either one may be the trigger source, and they may be displayed using the chop or alternate feature on most scopes.

In addition to displaying channel 1 and/or channel 2, most scopes provide the capability of displaying $1 + 2$ or $1 - 2$. Also, one or both of the channels may be capable of being displayed inverted (with its polarity reversed). $(1 - 2$ might not be provided, but $1 + 2$ with channel 2 inverted can achieve the same result.)

4.6 AC AND DC COUPLING

Each input can be selectively AC or DC coupled. DC coupling allows both DC and AC signals through, while AC coupling accepts only AC signals. Figure 4-15a shows a waveform containing both AC and DC. If the oscilloscope is DC coupled, then the waveform is displayed as drawn in Figure 4-15a. If the oscilloscope is AC coupled, then the DC portion of the waveform is blocked and only the AC portion is displayed as shown in Figure 4-15b.

The previous example seems very straightforward, but the issue of AC coupling may show up in other unexpected ways. Consider the pulse waveform in Figure 4-16a, shown as a DC coupled scope would display it. This waveform appears to be a typical AC waveform so one might think that it would be unaffected by coupling. However,

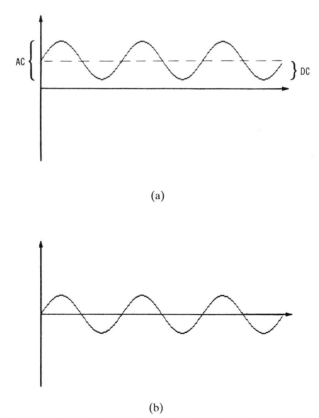

(a)

(b)

Figure 4-15 The effect of DC and AC coupling. (a) DC coupling causes the entire waveform to be displayed, including the DC portion. (b) AC coupling removes the DC portion of the signal.

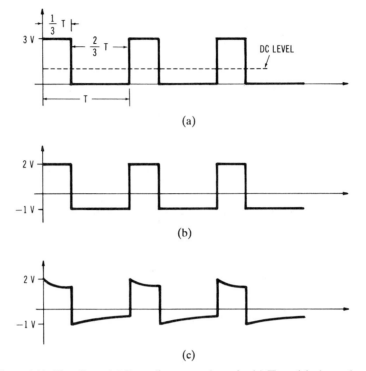

Figure 4-16 The effect of AC coupling on a pulse train. (a) The original waveform displayed with DC coupling. (b) The pulse train with DC component removed due to AC coupling. (c) AC coupling may cause voltage droop due to the loss of low frequencies.

when the scope is AC coupled, the display does change. The waveform shifts down by about one-third of its original zero-to-peak value (Figure 4-16b). The original waveform did have some DC present in it (remember, DC is just the average value of the waveform). The AC coupling removed the DC, leaving a waveform whose average value is zero. Notice that the waveform is not centered exactly on zero volts, since its duty cycle is 1/3. AC coupling may also cause voltage "droop" or "sag" in the waveform (Figure 4-16c), due to the loss of low frequencies.

Most oscilloscopes have a convenient means of grounding the input (usually a switch near the connector). This is symptomatic of one of the most confusing things in using an analog scope—where is zero volts on the display? The ground switch allows the user to quickly ground the input and observe the flat trace on the display which is now at zero volts. The line may then be set anywhere on the display that is convenient, using the scope's position controls. Knowing where zero is defined along with the volts per division selection determines the scale on the display.

Many scopes provide a BANDWIDTH LIMIT control which activates a fixed-frequency low-pass filter in the vertical amplifier. This has the effect of limiting the bandwidth of the scope (typically to about 20 MHz). Since bandwidth is such a desirable thing, it may seem odd to limit it intentionally. Figure 4-17a shows an oscilloscope display of a sine wave with a noticeable amount of high-frequency noise

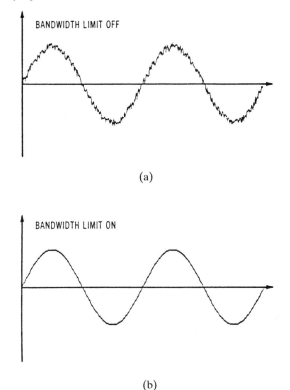

(a)

(b)

Figure 4-17   The effect of using the bandwidth limit control. (a) A noisy sine wave. (b) The same sine wave with the noise reduced by limiting the bandwidth.

riding on it. When the bandwidth limit control is switched on (Figure 4-17b), the high-frequency noise is eliminated, but the original sine wave remains uncorrupted. Of course, this works only when the interfering noise is mostly outside the bandwidth of the filter and the desired signal is inside the filter bandwidth.

4.7 X–Y DISPLAY MODE

Most two-channel oscilloscopes have the ability to plot the voltage of one channel on the vertical (Y) axis and the voltage of the other channel on the horizontal (X) axis (Figure 4-18). This results in a voltage versus voltage display, usually called channel-versus-channel or X–Y mode. The timebase and triggering circuits are not used when operating in this manner.

This feature greatly enhances the usefulness of the oscilloscope. The horizontal axis is no longer limited to only time. Any other parameter that can be represented as a voltage can now be used as the X axis. More precisely, both the vertical and horizontal axes can be used to display any two parameters represented by voltages. For instance, if a current sensing resistor were used to convert a current into a voltage,

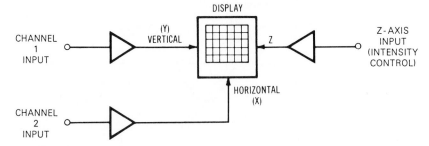

Figure 4-18 A simplified block diagram showing the X–Y display capability of the oscilloscope. The Z-axis input can be used to control the intensity of the plot at any given point.

a current could be plotted on the vertical axis while another voltage is plotted along the horizontal axis. This X–Y application, and others, are discussed in Chapter 5.

4.7.1 Z-Axis Input

The Z-axis input (also known as the INTENSITY MODULATION INPUT) provides a means for controlling the intensity of the display while in X–Y operation. The name "Z axis" comes from the fact that in addition to the X axis and Y axis information, the intensity of the display can be varied to provide an additional "axis" of information— the Z axis. If a positive voltage (typically several volts) is applied to the Z-axis input, the trace is blanked (no intensity). If a negative voltage is applied to the Z-axis input, then the trace has full intensity. Voltages in between the two extremes produce less than full intensity (but not blanked). The actual voltages and even the polarity of the Z-axis input varies depending on the model of instrument. Given the proper Z-axis control signals, different sections of the trace can have different intensities. This is useful for highlighting a particular point on a display or for turning the trace off to start the plot over at a particular point (without having a trace drawn to that point).

4.8 HIGH-IMPEDANCE INPUTS

The typical oscilloscope has a high-impedance input so that the circuit under test is not loaded significantly. The input can be modeled by a 1-MΩ resistor in parallel with a capacitance (Figure 4-19). The value of the capacitance depends on the particular model of oscilloscope, but is generally in the range of 7 to 30 pF. The magnitude of a typical input impedance is plotted in Figure 4-20. At low frequencies,

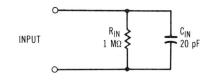

Figure 4-19 Circuit model for the input of an oscilloscope. The input capacitance depends on the particular instrument model, but is usually in the range of 7 to 30 pF.

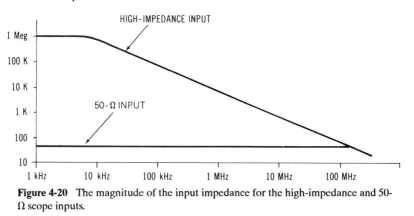

Figure 4-20 The magnitude of the input impedance for the high-impedance and 50-Ω scope inputs.

the capacitance acts like an open circuit and the impedance consists only of the 1-MΩ resistor. At about 8 kHz, the capacitor's impedance becomes significant as it just equals the 1-MΩ resistor impedance. The impedance of the parallel combination continues to decrease gradually for higher frequencies. Although the input impedance is very high at low frequencies, it will tend to load the circuit being measured as the frequency increases. Remember, a 1-MΩ input is not 1 MΩ at high frequencies.

4.9 50-Ω INPUTS

Some oscilloscopes offer a second type of input having a 50-Ω input impedance (Figure 4-21). It is often the same connector as the high impedance input, with a switch selecting between the two. This may be implemented by placing a 50-Ω resistor in parallel with the 1-MΩ input. Since the 1 MΩ is much larger than 50 ohms, the effective parallel impedance is approximately 50 ohms. If a scope does not have the 50-Ω input built in, an appropriate load can be placed in parallel with the high-impedance input to produce the same result. The input impedance is modeled as a single 50-Ω resistor, with the input capacitance in parallel. In a 50-Ω system, the capacitive effect is less critical, resulting in a wider measurement bandwidth.[1] Figure 4-20 shows that even though the 50-Ω input impedance starts out much smaller than the high-impedance input, it remains constant out to a higher frequency. The major disadvantage of the 50-Ω input is that it is too low a load resistance for many circuits. (For these cases, very-low-capacitance active probes which are designed to drive the 50-Ω input are used to provide minimal circuit loading and greater overall bandwidth.) Of course, the 50-Ω input is especially convenient for systems that have an inherent 50-Ω impedance.

---

[1]Oscilloscopes optimized for high frequency measurements may have compensation circuits which remove the effect of the parallel capacitance.

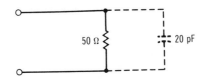

Figure 4-21 The circuit model for the 50-Ω scope input. The capacitance is often neglected in a 50-Ω system.

4.10 OSCILLOSCOPE SPECIFICATIONS

A set of typical scope specifications is shown in Table 4-1. Bandwidth and rise time have already been covered in the general discussion in Chapter 1. The DEFLEC-TION FACTOR tells what vertical sensitivity settings are available. In this example, settings between 5 mV and 5 V per division are available in a 1–2–5 sequence (5 mV, 10 mV, 20 mV, 50 mV, etc.). Included with the deflection factor is a percent error (± 3%) which defines the fundamental accuracy of the instrument. Similarly, the time per division settings on the horizontal axis are from 2 nsec to 0.5 sec per division. Again, an error specification which defines the accuracy of the time scale is included (± 3%).

TABLE 4-1. SPECIFICATIONS OF A TYPICAL 100-MHz ANALOG OSCILLOSCOPE.

Bandwidth (3 dB): 100 MHz
Rise time: 3.5 nsec
Input impedance: 1 MΩ in parallel with 22 pF
DC gain accuracy: ± 3%
Deflection factor: 5 mV/division to 5 V/division, ± 3%,
                in a 1–2–5 sequence
Sweep time: 2 nsec/division to 0.5 sec/division, ± 3%,
                in a 1–2–5 sequence

Oscilloscopes are normally specified for accuracy only at DC. It is assumed that the frequency response of the scope is flat at low frequencies and rolls off at higher frequencies. To achieve good pulse response (i.e., very little overshoot and ringing), the scope's response *must* roll off gradually with frequency. At its specified bandwidth, the scope's response is approximately 3 dB down from the DC response. Thus, a 1-V sine wave with a frequency near the scope's bandwidth will be measured as 0.707 V.

4.11 ANALOG OSCILLOSCOPE BLOCK DIAGRAM

The block diagrams shown previously highlighted the configuration for a particular type of oscilloscope operation. A block diagram incorporating all the features previously discussed is shown in Figure 4-22. The various switches allow the scope to be configured such that a wide variety of measurement functions can be performed. Both

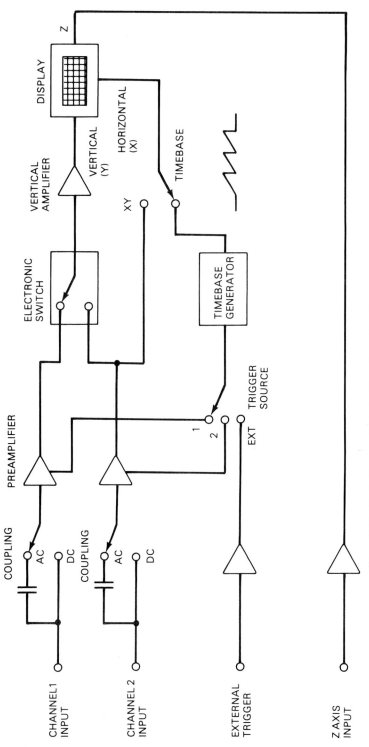

Figure 4-22 Block diagram of a typical two-channel oscilloscope.

input channels can be configured for DC or AC coupling, the electronic switch allows both alternate and chop modes for two-channel operation, and the display can be set up for either timebase or X–Y operation.

4.12 SCOPESMANSHIP

An oscilloscope is a fairly complex instrument, especially when compared with a simpler instrument such as a voltmeter. The large number of switches and knobs on the front panel can be intimidating to the novice user. A few comments are needed to help the first-time user get started.

After carefully reading the manufacturer's operating manual, the best way to get started with a measurement is to put the oscilloscope into a known state which will at least get something on the display. Many scopes have included a feature called AUTOSCALE or AUTOSETUP which automatically evaluates the waveform, chooses an appropriate trigger condition, and selects reasonable horizontal and vertical scales. Autoscale may not produce the optimum setup for the user, but it will generally get the signal on screen so the user can manually zero in on the right measurement. Pushing this one button can prevent a lot of user frustration. Assuming that this feature is not available, a suggested starting point for an oscilloscope measurement is listed in Table 4-2. This can only be a start as each measurement is somewhat different.

It is hoped that a trace will appear on the display after the oscilloscope is set up and the scope settings can be optimized for the particular measurement. If there is no trace at all, then things like the power switch (yes, the power switch!) and intensity control should be checked. Perhaps the waveform is just off screen because it is much larger than the volts/division setting will allow on screen. Try grounding the input—a horizontal line corresponding to zero volts should appear. Some analog scopes have a BEAM FINDER button which, when pushed, gives the user some idea where an off-screen trace is hiding.

If the trace is on screen, but is not stable, then the triggering controls should be adjusted. Try tweaking the trigger level to make the waveform stable. The slope and

TABLE 4-2. SUGGESTED STARTING POINT FOR OSCILLOSCOPE MEASUREMENTS.

Vertical amplifier
  Input coupling = DC
  Volts/division = 1 volt or (expected V_{0-P})/4
Timebase
  Timebase operation
  Time/division = 1 msec or 1/ (4 × expected frequency)
  Automatic sweep
Trigger
  Internal trigger (viewed channel)
  Trigger level = 0 volts
  Trigger slope = positive
  Trigger coupling = DC

trigger coupling may also be helpful. If the display is stable but is scaled improperly, adjust the time/division and/or the volts/division knobs.

Probably the best advice for operating an oscilloscope is *carefully try something*. The two approaches that do not work are (1) just sitting in front of the instrument, staring at it, and (2) twisting every knob until all the controls are guaranteed to be in the wrong position. Instead, the user should make an educated guess as to what control might fix the problem and try it. If it does not improve the situation, the control should probably be returned to the original setting. Try not to get the oscilloscope so fouled up that only a seasoned technician can straighten it out. If in doubt, revert back to the suggested starting point.

4.13 DIGITAL OSCILLOSCOPES

Conceptually, analog and digital (or DIGITIZING) oscilloscopes do the same thing—they display voltage waveforms. The analog scope uses traditional circuit techniques to display the voltage waveform on a CRT. A digital scope, in contrast, converts the original analog signal to digital form (a series of binary numbers) which can then be displayed or stored in memory. This means that a digital scope is inherently a storage scope, because the waveform is stored digitally. (Digital scopes are often called DIGITAL STORAGE OSCILLOSCOPES or DSOs.) Contrast this with the analog scope, where the waveform is a short-lived voltage waveform. When the input disappears, the displayed waveform disappears. There are techniques for storing waveforms on an analog display, but they generally are expensive and temperamental and have finite storage time. The ability to store waveforms is especially important when capturing one-time events. Without wave-form storage, the waveform is plotted across the display and then abruptly disap-pears. With storage, the waveform remains on the screen so that the user can carefully analyze the data. Two typical digital oscilloscopes are shown in Figures 4-23 and 4-24.

The block diagram of a digital oscilloscope is shown in Figure 4-25. The digitizing oscilloscope has input circuitry which is similar to an analog scope. The preamplifier output signal is sampled using either a track-and-hold or a sample-and-hold circuit and digitized by an analog-to-digital converter (ADC). After the ADC, the signal is in digital form. The sample clock drives the ADC, the sampler, and the ADC memory, controlling when samples are acquired in time.

In front of the sampler, the circuitry must support the full bandwidth of the scope, but after the ADC the bandwidth requirements are much lower. In particular, since the waveform is stored in memory, the bandwidth of the display only needs to be fast enough to refresh the screen (from memory) without visible flicker. Therefore, a lower bandwidth display can be used to display the waveform. (Compare this with an analog scope where the entire system, including the display, must support the full instrument bandwidth.)

In Figure 4-25, a sampler circuit is shown in front of the ADC, which is the general case. Some digital oscilloscopes use ADCs that are fast enough to capture the analog signal without the need for a sampler circuit.

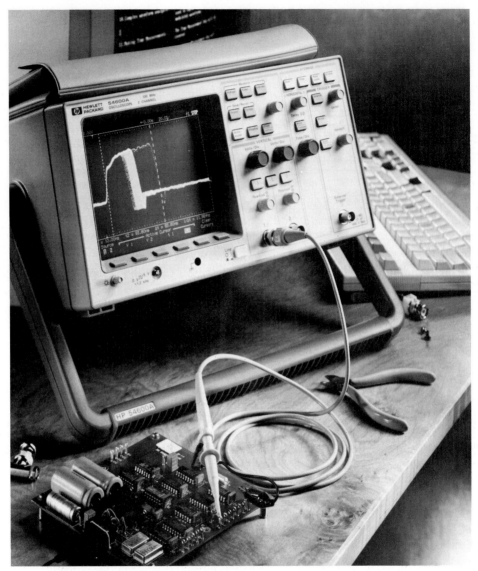

Figure 4-23 A 100-MHz bandwidth digitizing oscilloscope shown connected to a circuit using a 10:1 probe. (Photo courtesy of Hewlett-Packard Company.)

The function of the timebase circuit is significantly different in a digital scope. The timebase does not produce the ramp voltage as an analog timebase does. Instead, the timebase circuit uses a crystal oscillator to measure the time difference between the trigger signal and the sample clock. With this time difference known, the processor can determine where to put the waveform samples on the display. The processor can be a general-purpose microprocessor or dedicated digital logic that writes the sampled waveform data to the display.

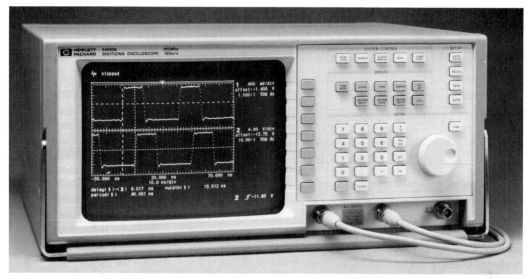

Figure 4-24   A digitizing oscilloscope with a 1-Gsa/sec sample rate. (Photo courtesy of Hewlett-Packard Company.)

4.13.1 Vertical Resolution

The ADC in a digital scope converts the input voltage to a binary number. This process results in QUANTIZATION of the signal. That is, a continuously varying input waveform is converted into a series of discrete values.

A binary number having n bits can represent 2^n different values or codes. For a voltage range of V_{RANGE}, the voltage resolution of an n-bit converter is given by

$$V_{RES} = \frac{V_{RANGE}}{2^n}$$

Example 4-1.

What is the voltage resolution of a digital oscilloscope with an 8-bit ADC when the sensitivity is 200 mV/div? (Assume that the ADC range is spread over eight vertical divisions.)

The voltage range covered by the ADC is 8×200 mV $= 1.6$ volts, so the resolution is given by

$$V_{RES} = \frac{1.6}{2^8} = 6.3 \text{ mV}$$

The voltage resolution can also be expressed as a percentage of the full-scale voltage range. Table 4-3 shows the vertical resolution for various numbers of ADC bits.

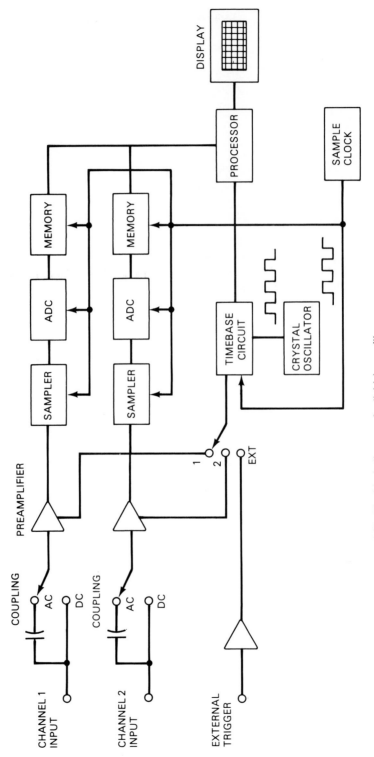

Figure 4-25 The block diagram of a digitizing oscilloscope.

TABLE 4-3   ANALOG-TO-DIGITAL
CONVERTER BITS AND VERTICAL
RESOLUTION

Bits	Resolution
6	1.56%
7	0.78%
8	0.39%
9	0.20%
10	0.098%
11	0.049%
12	0.024%

4.13.2   Sample Rate

The SAMPLING THEOREM states that for a baseband signal to be faithfully reproduced in sampled form, the sample rate must be greater than twice the highest frequency present in the signal.

$$f_s > 2\,BW$$

where

f_s = sample rate
BW = bandwidth of the signal

The minimum sample rate is known as the NYQUIST RATE. This relationship guarantees that slightly more than two samples per cycle are acquired on a sine wave having a frequency right at the bandwidth limit. This equation defines the minimum theoretical sample rate. In practice, a higher sample rate is required to represent the signal accurately.

The sample rate and bandwidth specifications on a digital scope tend to be confused. Does a 100-MHz digital scope have a sample rate of 100 MHz or a bandwidth of 100 MHz? To help alleviate this problem, digital scope manufacturers have settled on specifying the sample rate in samples per second and the bandwidth in hertz. More typically, the sample rate is in megasamples per second (MSa/sec) and the bandwidth is in megahertz (MHz).

Example 4-2.

What is the minimum sample rate required by the sampling theorem to represent completely a signal having a 100-MHz bandwidth?

The sampling theorem requires

$f_s > 2\,BW$
$f_s > 2 \times 100$ MHz = 200 MSa/sec

The sample rate f_s must be greater than 200 MSa/sec.

This still leaves the user with the question of how do sample rate and bandwidth relate in a practical digital scope? Or how much sample rate is required for a particular bandwidth? As we shall see, the answer depends on the sampling technique used in the oscilloscope.

4.13.3 Real-time Sampling

Real-time sampling is the most obvious and intuitive type of sampling. This type of sampling simply acquires sample points uniformly spaced in time (Figure 4-26). All the sample points are acquired in response to one occurrence of the oscilloscope trigger event. The major advantage of this technique is that a one-time transient (or SINGLE-SHOT EVENT) can be acquired. The disadvantage is that the analog-to-digital converter must operate above the Nyquist rate to capture the signal correctly.

Real-time sampling can provide waveform information in front of the trigger event. This pretrigger information is obtained by letting the sampling system continuously acquire data which is retained in digital memory and displayed when a trigger event finally occurs. Pretrigger viewing is a powerful advantage of the digitizing scope architecture, since the user can trigger on some failure condition (or other event) and then look back in time to determine what caused the failure. (The time before the trigger event is often referred to as NEGATIVE TIME.)

The memory depth is important in real-time sampling, since it is often desirable to acquire long time records at a high sample rate. For a given sample rate, deeper memory allows longer time records to be acquired. Of course, the sample rate can also be decreased, allowing longer events to be digitized, but at the expense of lower bandwidth and poorer timing resolution.

Oscilloscopes that are optimized for real-time sampling of single-shot events use interpolation between sample points to reconstruct the waveform. This interpolation algorithm usually takes the form of a digital filter following the ADC (Figure 4-27). This reconstruction filter fills in between sample points when the sample rate is operating near the Nyquist rate. The closer the sample rate is to twice the bandwidth, the steeper the reconstruction filter must be.

The single-shot bandwidth of a digital scope is defined as the maximum bandwidth that can be captured in a single-shot measurement. The relationship between the single-shot bandwidth of a real-time scope and the sample rate is given by

$$BW_S = \frac{f_s}{k_R}$$

where

BW_S = the single-shot bandwidth
f_s = the sample rate
k_R = a factor depending on the reconstruction technique

The factor k_R can theoretically approach 2 (according to the sampling theorem), but is typically 2.5 or greater. With $k_R = 2.5$, reconstruction is suitable only for sine

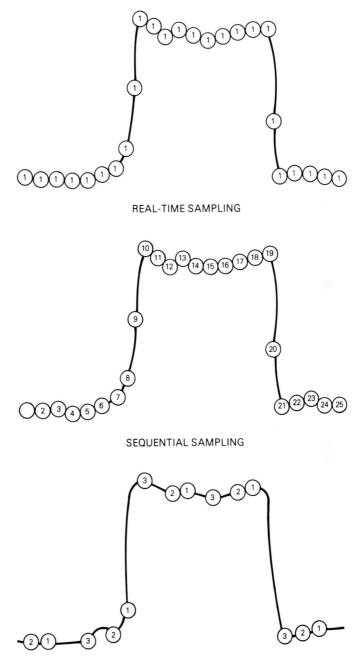

REAL-TIME SAMPLING

SEQUENTIAL SAMPLING

RANDOM REPETITIVE SAMPLING

Figure 4-26   Digitizing oscilloscope sampling techniques. The numbers on the samples indicate which trigger event the sample was acquired on. (Courtesy of Hewlett-Packard Company.)

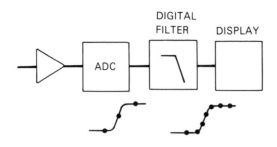

Figure 4-27 A digital reconstruction filter provides interpolation between sample points in a real-time scope.

wave signals, since the reconstruction filter shape must be steep, which causes ringing
on pulsed signals. For a typical transient phenomenon such as a pulse, a factor of k_R
= 4 allows a reconstruction filter with a good transient response to be used. With k_R
= 4, a typical oscilloscope having a sample rate of 1 GSa/sec has a single-shot
bandwidth of 250 MHz. In cases where no reconstruction is used, k_R = 10 is required,
giving 10 samples per period of the highest-frequency sine wave.

Example 4-3.

An oscilloscope uses real-time sampling with digital reconstruction using k_R = 4. If the
sample rate of the scope is 400 MSa/sec, what is the single-shot bandwidth of the scope?
If no reconstruction is used, what is the single-shot bandwidth?

The single-shot bandwidth is given by

$$BW_S = \frac{f_s}{k_R}$$

For the k_R = 4 case,

$$BW_S = \frac{400 \times 10^6}{4} = 100 \text{ MHz}$$

With no reconstruction used, k_R = 10 so

$$BW_S = \frac{400 \times 10^6}{10} = 40 \text{ MHz}$$

4.13.4 Repetitive Sampling

Repetitive sampling (also called EQUIVALENT-TIME SAMPLING) includes two
different sampling techniques—sequential sampling and random repetitive sampling.
These two techniques are shown in Figure 4-26. Either sampling technique can be
used if the following two conditions are met.

1. The waveform must be repetitive.
2. A stable trigger event must be available.[2]

---

[2]Most oscilloscope measurements require a stable trigger. This requirement is mentioned here to
emphasize how repetitive sampling uses the trigger event to keep track of where the samples occur.

Since the waveform is repetitive, the scope can acquire the waveform samples over many periods of the waveform. A stable trigger event is required to arrange these samples on the display properly.

Sequential sampling operates by acquiring a single sample on each trigger event (see Figure 4-26). When a trigger occurs, a sample is taken at a precisely controlled time after the trigger. On each trigger, this delay time is increased so that all portions of the waveform are eventually sampled. As long as the time delay from the trigger point to the acquired sample is precisely controlled, the waveform is accurately reproduced. The chief advantage of this technique is that a much slower ADC can be used. (The sampler must still sample the signal very quickly, to minimize time jitter in the sample.) This technique has been used to implement a microwave scope which has a bandwidth of 50 GHz using an ADC with a sample rate of only 10 kSa/sec. Real-time sampling of this bandwidth would require an ADC sample rate of well over 100 GSa/sec, which is not currently realizable.

Since all the samples occur after the trigger event, sequential sampling provides no pretrigger (negative time) information. Random repetitive sampling overcomes the pretrigger limitation of sequential sampling. By sampling randomly (even before a trigger event occurs), some of the acquired samples end up occurring in front of the trigger. The timebase circuit of the scope must measure the time between the trigger event and the digitized samples, so that the samples can be placed properly on the display. As shown in Figure 4-26, random repetitive sampling may acquire multiple samples per trigger event, depending on the scope's time/division setting.

The performance of a random repetitive scope is not limited so much by its ADC speed, but by the ability of the scope to acquire a sample precisely with minimum time jitter and then place that sample point in time relative to the trigger event. Figure 4-28 shows a high-resolution measurement of a 30-nsec-wide pulse using a random repetitive scope with a 10-MSa/sec sample rate. Note that the period of the sample rate is 100 nsec, which is several times wider than the pulse, but the waveform is displayed with 100-psec timing resolution.

It may appear that the sampling theorem is being violated, in that a slow ADC is being used to sample a high-frequency signal. Sometimes the concept of effective sample rate is introduced to alleviate the concern.

$$f_{EFF} = \frac{1}{T_{EFF}}$$

where

f_{EFF} = the effective sample rate
T_{EFF} = the effective sample period, which is equal to the time between samples after the waveform is acquired

T_{EFF} is not dependent on the sample rate of the ADC, but is dependent on how precisely the samples can be placed in time relative to the trigger event. After the

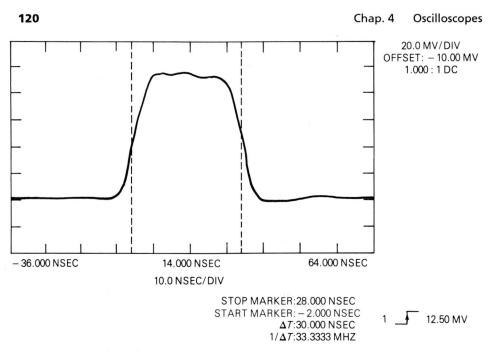

20.0 MV/DIV
OFFSET: – 10.00 MV
1.000 : 1 DC

– 36.000 NSEC   14.000 NSEC   64.000 NSEC

10.0 NSEC/DIV

STOP MARKER:28.000 NSEC
START MARKER: – 2.000 NSEC
ΔT:30.000 NSEC   1   ⌐_ 12.50 MV
1/ΔT:33.3333 MHZ

Figure 4-28   A 30-nsec pulse is accurately measured even though the random repetitive sample rate of 10 MSa/sec has a period wider than the pulse.

entire waveform is acquired, T_{EFF} is the time between adjacent sample points.
Effective sample rate is really a measure of the timing resolution of the instrument.
In a repetitive scope, the effective sample rate will be much higher than the actual
ADC sample rate and will satisfy the sampling theorem. For the measurement shown
in Figure 4-28, T_{EFF} is 100 psec (the timing resolution of the scope), so $f_{EFF} = 10$
GSa/sec.

Since random repetitive sampling provides pretrigger information, it has
largely displaced sequential sampling, except at microwave frequencies. At microwave frequencies, the time/division setting on the scope can be very small, causing
the window of time that is viewed on the display to also be very small (perhaps
100 ps). The probability of a randomly acquired sample falling into the desired time
window is so small that random repetitive sampling would take a long time to
acquire the entire waveform. Alternatively, sequential sampling forces the sample
points to occur within the desired time window so the entire waveform can be
acquired quickly.

Many digitizing scopes combine both real-time and repetitive sampling techniques, using real-time sampling for lower-frequency bandwidths and repetitive
sampling on the higher frequencies. Also, random repetitive scopes have some
inherent single-shot or real-time bandwidth. Since reconstruction techniques are not
typically used in a strictly random repetitive scope, the single-shot bandwidth is
usually specified as $f_s/10$.

4.13.5 Digital Acquisition and Display Techniques

The architecture of a digital scope opens up some possibilities for processing the acquired waveform that are not possible with analog scopes. With the waveform in digital form, the oscilloscope's processor can operate on the data before writing it to the display.

The microprocessor can perform a digital persistence algorithm, which produces a display similar to the traditional analog storage scope. When a particular sample is acquired, it is displayed for a set length of time, called the display time. If that same voltage/time location is not reacquired within the display time, it will be erased from the display. With a changing waveform, this causes the older waveform data to fade out while the new waveform data remains.

Of particular interest is the case where the persistence is infinite, which means the acquired samples are displayed indefinitely. INFINITE PERSISTENCE provides for worst case characterization of signals. Since persistence is implemented digitally, there is no problem with the display blooming as with analog storage scopes. One use of infinite persistence is the measurement of worst case timing jitter on a signal (Figure 4-29). Other applications of infinite persistence are eye diagrams (as used in telecommunication channels) and detection of metastable states in logic design.

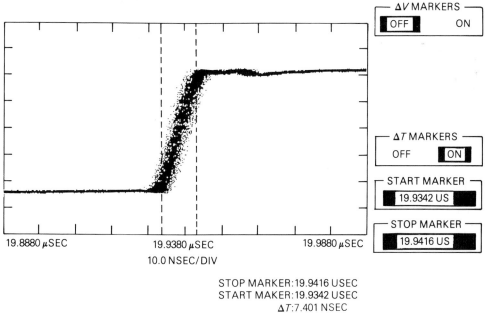

Figure 4-29. The time jitter present in a signal is measured using infinite persistence. The scope is configured to trigger on one edge of the waveform while viewing a later edge.

One problem with the use of infinite persistence is that the signal can get lost among the older waveform information. A slight variation on the display technique, called AUTOSTORE, solves this problem by displaying the most recent waveform information in full intensity while older waveform samples are displayed at half intensity. This allows the oscilloscope user to view changes in the waveform while simultaneously keeping track of worst case excursions. A typical jitter measurement using autostore is shown in Figure 4-30.

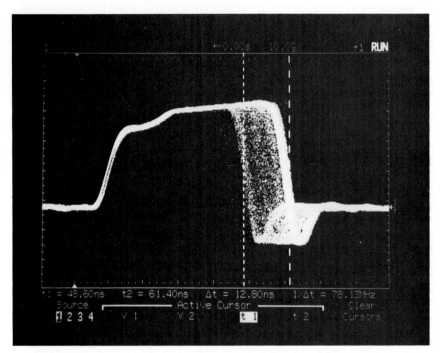

Figure 4-30 Jitter measurement using autostore function. (Photo courtesy of Hewlett-Packard Company.)

In a digitizing oscilloscope is performed by averaging together sample points associated with the same point in time, but from different acquisitions. Averaging done is this fashion reduces the noise[3] present in the measurement without bandwidth or rise time degradation (Figure 4-31). (In the analog scope case, bandwidth limit may be used, which does reduce the bandwidth and increase rise time.)

Averaging has the side effect of slowing down the display's responsiveness. Usually the amount of averaging is selectable, with more averaging causing a more sluggish display. A more positive effect of averaging is increased vertical resolution as multiple waveform samples are averaged together.

---

[3]For averaging to remove the noise, the noise must be asynchronous to the trigger signal. Noise which is synchronous to the trigger will remain in the measurement.

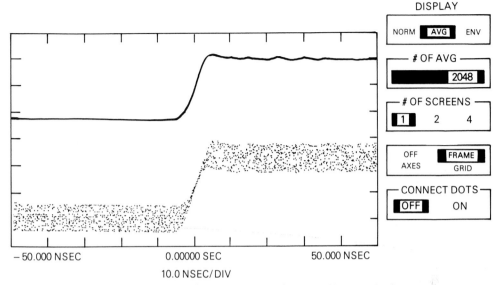

Figure 4-31   The noise present in the bottom trace is removed by averaging in the top trace. Note that there is no rise time (bandwidth) degradation.

4.14 ANALOG SCOPE VERSUS DIGITAL SCOPE

While the analog scope and the digital scope perform basically the same function, there are some differences which separate the two scope types in practical use.

The analog oscilloscope has a highly interactive, high-resolution display which can show even small variations in waveform amplitude and intensity. Ideally, the vertical and horizontal resolution of the analog scope are infinitely small. There are no quantization effects to limit the resolution as in the case of the digital scope. In reality, the minimum dot size of the display and its trace width will limit the ultimate resolution. On fast-sweep speeds (especially with low trigger rates), the analog scope trace loses intensity quite dramatically, often forcing the user to block out ambient light with a viewing hood. The analog oscilloscope historically has enjoyed a cost advantage over the equivalent bandwidth digital scope. However, as digital technology has improved, the analog cost advantage has eroded significantly.

The digital scope has the advantage of being able to capture and display single-shot events and the ability to freeze the display on repetitive waveforms. The timebase of the digitizing scope is controlled by a crystal oscillator, which typically has a frequency accuracy of 0.005 percent. Since the horizontal axis timing information is derived from this oscillator, the timebase linearity and accuracy are much better than that of an analog scope.

As discussed earlier in the chapter, the vertical position of an analog scope display is not calibrated. Usually the scope user must manually set the location of zero volts by grounding the input and adjusting the vertical position. Digital scopes usually maintain a calibrated vertical position and indicate where ground is on the display.

The digitizing architecture inherently produces the waveform in digital form, which provides easy implementation of a number of features. Waveform storage, automatic measurements, and waveform transfers to printer or plotter are common features on a digitizing scope. Most digitizing scopes can automatically compute and display the peak-to-peak voltage, the rise time, period, and frequency, of a waveform rather than having the user manually do it. (Some analog scopes offer these automatic measurements by adding dedicated circuitry to perform the measurement, causing the cost to increase.) A good digital scope can even find the RMS voltage of the waveform by performing a root-mean-square calculation on the data. Digital storage of the displayed waveform is especially useful on the slow time ranges and on low-repetition-rate signals because display flicker (as with an analog scope) is avoided. Digital display storage is also useful on fast, infrequent signals that would have very low intensity on an analog scope. Computer interfaces (such as the IEEE-488 and RS-232 interfaces) provide access to the waveform data in automatic test applications.

4.15 OSCILLOSCOPE PROBES

The high-impedance input of the oscilloscope can be connected directly to the circuit under test using a simple cable. It is highly recommended that any cable used be shielded to minimize noise pickup. However, for maximum performance an oscilloscope probe matched to the input of the scope should be used.

4.15.1 1× Probes

1× probes, also known as 1:1 (one-to-one) probes, simply connect the high-impedance input of the oscilloscope to the circuit being measured. They are designed for minimum loss and easy connection, but otherwise they are equivalent to using a cable to connect the scope. Figure 4-32 shows the circuit diagram for a high-impedance scope input connected to a circuit under test. The circuit under test is modeled as a voltage source with a series resistor. The 1× probe (or cable) will introduce a significant amount of capacitance which appears in parallel with the input of the scope. A 1× probe may have around 40 to 60 pF of capacitance, which is usually larger than the oscilloscope input capacitance.

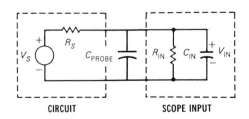

Figure 4-32 The high-impedance input is connected to a circuit using a 1 × probe.

4.15.2 Loading Effects

The impedance of the circuit and the input impedance of the oscilloscope together produce a low-pass filter. For very low frequencies, the capacitor acts as an open circuit and has little or no effect on the measurement. For high frequencies, the capacitor's impedance becomes significant and loads down the voltage seen by the oscilloscope. Figure 4-33 shows this effect in the frequency domain. If the input is a sine wave, the amplitude tends to decrease with increasing frequency and the phase is shifted.

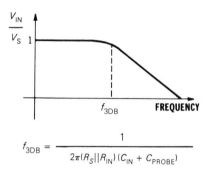

$$f_{3DB} = \frac{1}{2\pi (R_S || R_{IN})(C_{IN} + C_{PROBE})}$$

Figure 4-33 In the frequency domain, the response of the 1× probe rolls off at the higher frequencies.

The loading also affects the oscilloscope's response to a step change in voltage. The loading due to the input impedance of the scope (and the probe capacitance) can be broken into two parts: RESISTIVE LOADING and CAPACITIVE LOADING. Figure 4-34 shows the probe and scope input loading broken into resistive and capacitive loading, which can be analyzed independently. The resistive loading is due

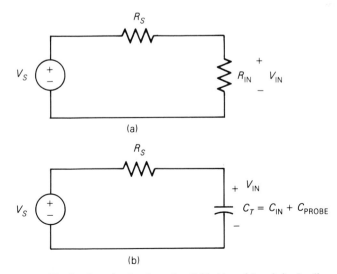

Figure 4-34 The loading of a circuit can be divided into (a) resistive loading and (b) capacitive loading.

entirely to the input resistance of the scope, while the capacitive loading is due to the probe capacitance combined with the scope input capacitance.

The resistive loading circuit of Figure 4-34 is another example of the voltage divider circuit. Thus, the voltage delivered to the scope input, V_{IN}, is a replica of V_S but with reduced amplitude. For a voltage step from zero to V_{MAX} at time $t = 0$,

$$v_{IN}(t) = V_{MAX} \frac{R_{IN}}{R_{IN} + R_S} \qquad \text{for } t > 0$$

The effect of capacitive loading is more complex and results in an exponential response in the voltage. For a V_S voltage step which goes from zero volts to V_{MAX} volts, V_{IN} obeys the following equation. (See Appendix D for the complete analysis.)

$$v_{IN}(t) = V_{MAX} [1 - e^{-t/(R_S C_T)}] \qquad \text{for } t > 0$$

The step responses due to the two loading effects are shown in Figure 4-35. The resistive loading changes the size of the voltage step, but does not change the waveform shape. Capacitive loading slows down the rise time of the step but eventually settles out to the same final value as the ideal response. As shown in Chapter 1, the bandwidth and rise time of a system are inversely related. Since the bandwidth of the instrument is effectively being decreased, the rise and fall times of pulse inputs will be increased.

The circuit model used for this analysis may not be accurate for all types of practical circuits. The output resistance (drive capability) of digital circuits may vary with the output voltage and cause the loading effect to be different. Even though this model is not 100 percent accurate for such a circuit, the basic principle of resistive and capacitive loading still applies. This means that load capacitance will slow down the rise time of the signal while resistive loading will tend to change the output amplitude. Increased rise time in a digital circuit is translated into increased delay when the signal

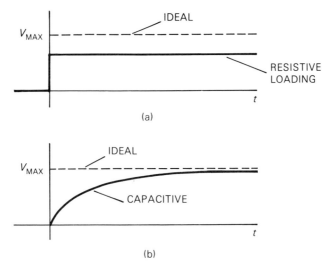

Figure 4-35 Resistive loading (a) changes the voltage level of a step while capacitive loading (b) causes an exponential response.

reaches the next logic gate. This is because it will take longer for the signal to rise to the logic threshold, causing the next gate to switch later. The 1-MΩ input impedance of the typical oscilloscope is large enough to prevent resistive loading of most digital circuits, but the capacitive loading of a 1:1 probe will introduce significant delay into the signal.

4.15.3 10 $\times$ Probes

10$\times$ probes (also called 10:1 probes, divider probes, or attenuating probes) have a resistor and capacitor (in parallel) inserted into the probe. Figure 4-36 shows the circuit for the 10$\times$ probe connected to a high-impedance input of an oscilloscope. If $R_1C_1 = R_2C_2$, then this circuit has the amazing result that the effect of both capacitors exactly cancel. In practice, this condition may not be met exactly but can be approximated. The capacitor is usually made adjustable and can be tweaked for a near perfect match. Under these conditions, the relationship of V_S to V_{IN} is

$$V_{IN} = V_S\left(\frac{R_2}{R_1 + R_2}\right)$$

which should be reminiscent of the voltage divider equation. R_2 is the input resistance of the scope's high input impedance (1 MΩ) and $R_1 = 9R_2$. From the previous equation, this results in

$$V_{IN} = \left(\frac{1}{10}\right)V_S$$

So the net result is a probe and scope input combination that has a much wider bandwidth than the 1 $\times$ probe, due to the effective cancellation of the two capacitors. The penalty that is incurred is the loss of voltage. The oscilloscope now sees only one-tenth of the original voltage (hence the name 10$\times$ probe). Also notice that the circuit being measured sees a load impedance of $R_1 + R_2 = 10$ MΩ, which is much

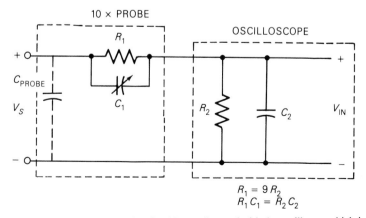

Figure 4-36 Circuit diagram showing 10 $\times$ probe used with the oscilloscope high-impedance input. The effect of the capacitors cancel when C_1 is adjusted properly.

higher than with the 1× probe. Some probes are designed to be conveniently switched between 1× and 10× operation.

With a 10× probe, both the resistive and capacitive loading effects are reduced (relative to a 1× probe).[4] Although the input capacitance of the scope is ideally canceled, there is a remaining capacitance due to the probe, C_{PROBE}. This capacitance, which is specified by the manufacturer, will load the circuit under test.

The factor of 10 loss in voltage is not a problem as long as the voltage that is being measured is not so small that dividing it by 10 makes it unreadable by the scope. This means that the scope's sensitivity and the signal voltage may be factors in deciding whether to use a 10× probe. On most oscilloscopes, the user must remember that a 10× probe is being used and must multiply the resulting measurements by a factor of 10. This is a nuisance so some scopes include two scale markings: one valid for a 1× probe and the other valid for a 10× probe. Other scopes have gone one step farther and automatically adjust the readings by the correct amount when an attenuating probe is used.

4.15.4 Other Attenuating Probes

Other types of attenuating or divider probes are available, including 50:1 and 100:1 probes. The general principles of these probes are the same as the 10× divider probe: voltage level and bandwidth are traded off. To obtain wider bandwidth, more loss is incurred in the probe and less voltage is supplied to the input of the scope. This may require a more sensitive scope for low-level measurements. Some divider probes use the scope's 50-Ω input instead of the 1-MΩ input.

4.16 PROBE COMPENSATION

To maximize the bandwidth of the attenuating probe, the probe capacitor must be adjusted precisely such that the input capacitance of the scope is canceled. This is accomplished by a procedure known as COMPENSATION.

The scope probe is connected to a square wave source called the CALIBRA-TOR which is built into the scope. The probe is then adjusted to make the square wave as square and flat topped as possible. Figures 4-37a and 4-37b show the oscilloscope display during compensation with an overcompensated and undercompensated probe. Figure 4-37c shows the display when the probe is properly compensated.

As discussed in Chapter 1, the square wave is a wide bandwidth signal rich in harmonics. If the probe is adjusted so that it measures a square wave with a minimum of waveform distortion, then the probe will be correctly compensated for wide

---

[4]Some 10× probes have a resistor across the probe input so that the resistive loading is 1 MΩ. These probes do not represent an improvement in resistive loading over the 1× probe, but they do have less capacitive loading.

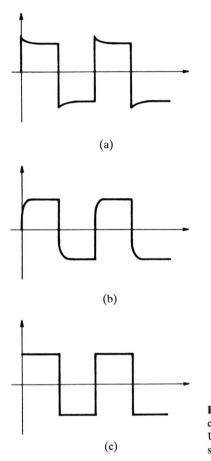

(a)

(b)

(c)

Figure 4-37 Examples of 10× probe compensation. (a) Overcompensated. (b) Undercompensated (c) Properly compensated.

bandwidth signals in general. This concept is also used for square wave testing of amplifiers as discussed in Chapter 5.

4.17 ACTIVE PROBES

So far, all the probes discussed have been simple passive circuits with no active components such as transistors or integrated circuits. In cases where extremely low capacitance is required for high-frequency measurements, an ACTIVE PROBE may be used. An active probe has a small amplifier built into it that is designed to have very little capacitance on its input. The output of the amplifier is usually matched to drive the 50-ohm input of the oscilloscope. This allows a length of 50-ohm cable to be used between probe and scope without any additional capacitive loading effects.

Table 4-4 summarizes the typical specifications of the various types of scope probes that have been discussed. Actual characteristics will vary according to manufacturer and model.

TABLE 4-4. TYPICAL SPECIFICATIONS OF OSCILLOSCOPE PROBES.

Probe Type	Frequency Range	Resistive Load	Capacitive Load
1×	DC to 5 MHz	1 MΩ	30 pF
10×	DC to 50 MHz	10 MΩ	10 pF
Active Probe	DC to 500 MHz	10 MΩ	2 pF
High Voltage	DC to 1 MHz	500 MΩ	3 pF

4.18 DIFFERENTIAL MEASUREMENTS

Most oscilloscope inputs are BNC[5] connectors with one side directly connected to instrument ground (which is in turn connected to the power line safety ground). This is the most practical and economical way to build the instrument. This is usually not a problem, since most voltage measurements are made with respect to ground. For some measurements, however, it is desirable to connect the scope input to arbitrary points in the circuit, including ones that are not grounded. Attaching a standard scope probe's ground to such a point will force that point to the scope ground, which is power line ground. Excessive current may flow, ruining the circuit or the measurement.

Some scopes have floating or differential inputs that allow both leads of the input to be connected away from ground. In this case, the grounding problem is avoided.

A two-channel scope with the ability to display channel $1 - 2$ (the difference between the two channels) can be used as a one channel floating input scope. The oscilloscope is set up to display $1 - 2$. Channel 1 is connected to the point in the circuit taken to be the more positive voltage. Channel 2 is connected to the other voltage point, and the oscilloscope ground is connected to the circuit ground. Thus, the scope displays the difference between the two voltage points with neither one required to be at ground.

A DIFFERENTIAL PROBE eliminates this problem by providing two scope probe inputs which can be floating relative to the scope's ground (Figure 4-38). The output voltage of the probe is the difference between the voltages on the two input terminals, allowing it to drive the ground-referenced input of an oscilloscope. The differential amplification is not perfect, and the error is specified in terms of COMMON MODE REJECTION RATIO (CMRR). To measure CMRR, both inputs are driven with the same signal. Ideally, the output (which is the difference between the two inputs) is always zero. But in a real probe there is some small output voltage.

$$CMRR = \frac{\text{input voltage (both inputs driven simultaneously)}}{\text{output voltage}}$$

[5]Bayonet Neill-Concelman, named for the two inventors of the connector Paul Neill and Carl Concelman.

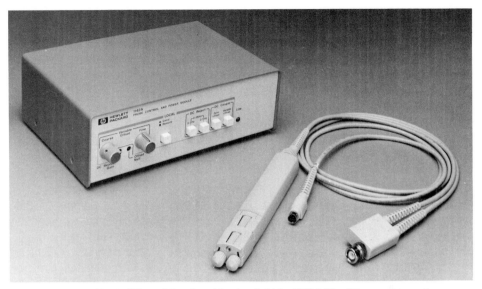

Figure 4-38 A differential probe with a bandwidth of 200 MHz. (Photo courtesy of Hewlett-Packard Company.)

Typically, the CMRR of a differential probe is best at low frequencies and degrades at higher frequencies. CMRR is often expressed in dB.

4.19 HIGH-VOLTAGE PROBES

Most conventional oscilloscope probes are rated to withstand operating voltages of around 450 volts (DC plus peak AC). For measuring higher voltages, a specialized HIGH-VOLTAGE PROBE must be used. High-voltage probes are typically rated from 5 kV to 30 kV (maximum) and have an attenuation factor of 1000. Such probes are physically large (compared to standard probes) to prevent high-voltage arcing. A typical high-voltage probe is shown in Figure 4-39.

4.20 CURRENT PROBES

Oscilloscopes are designed for a voltage input, but can be used to measure current using a CURRENT PROBE. A current probe has a set of jaws that encloses the wire that the measured current is flowing through. No electrical connection is needed. The circuit does not have to be broken or altered in any way, as the current probe measures whatever current is passing through its closed jaws.

Current probes generally use one of two technologies. The simplest uses the principle of a transformer, with one winding of the transformer being the measured wire. Since transformers work with only AC voltages and currents, current probes of this type do not measure direct current.

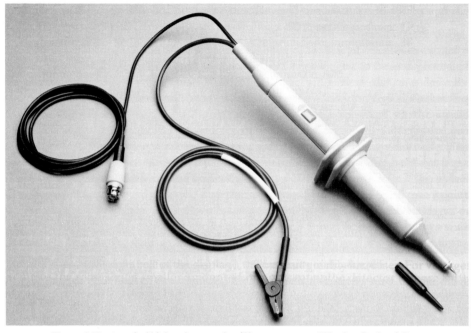

Figure 4-39   A typical high-voltage probe. (Photo courtesy of Hewlett-Packard Company.)

The other type of current probe works using the principle of the HALL EFFECT. The Hall effect produces an electric field in response to a current present in an applied magnetic field.[6] This technique requires the use of an external power supply, but does measure both alternating and direct current (AC and DC).

Since current probes measure the current enclosed by their jaws, several techniques can be used that are unique to the current probe. If the sensitivity of the probe and oscilloscope combination is too low for a particular measurement to be made, several turns of the current carrying wire can be inserted into the jaws. The probe will effectively have a larger current to measure (the original current times the number of turns). In a similar manner, the difference between two currents can be measured if the two wires in question are inserted, but with the currents flowing in opposite direction (the sum will be measured if the currents are flowing in the same direction). Of course, the physical size of the wires and the current probe will be a factor in determining how many wires can be inserted. Although the current does not require a direct electrical connection, it still removes energy from the circuit under test. Normally, this small amount of energy loss will not disturb the circuit, but can be a factor in some cases.

[6]For more information on the Hall effect, see Halliday and Resnick (1966).

REFERENCES

"Bandwidth and Sampling Rate in Digitizing Oscilloscopes." Hewlett-Packard Company, Application Note 344, Publication Number 5954-2631, April 1986, Palo Alto, CA.

HALLIDAY, DAVID, and ROBERT RESNICK, *Physics*, New York: John Wiley & Sons, Inc., 1966.

SCHLATER, RODNEY T. "Waveform Graphics for a 1-GHz Digitizing Oscilloscope." *Hewlett-Packard Journal*, April 1986.

STANLEY, WILLIAM D., GARY R. DOUGHERTY, and RAY DOUGHERTY, *Digital Signal Processing*, 2nd ed., Reston, VA: Reston Publishing Company, Inc., 1984.

"Understanding and Minimizing Probing Effects." Hewlett-Packard Company, Application Note 1210-2, Publication Number 5091-1800E, 1991, Palo Alto, CA.

"Voltage and Time Resolution in Digitizing Oscilloscopes." Hewlett-Packard Company, Application Note 348, November 1986, Palo Alto, CA.

WEDLOCK, BRUCE D., and JAMES K. ROBERGE. *Electronic Components and Measurements*. Englewood Cliffs, NJ: Prentice-Hall., Inc., 1969.

WITTE, ROBERT A. "Getting Scope Triggering in Your Sights." *Electronics Test*, August 1990.

WITTE, ROBERT A. "Understanding Digitizing Oscilloscopes." RF Technology Expo paper, February, 1989.

WITTE, ROBERT A. "Averaging Techniques Reduce Test Noise, Improve Accuracy." *Microwaves & RF*, February 1988.

chapter five

Oscilloscope Measurements

The oscilloscope produces a picture of the voltage waveform that is being measured. This allows the instrument user to think in terms of the actual waveform when making measurements. Therefore, determining zero-to-peak voltages, RMS voltages, and the like simply means interpreting the scope display using the theory in Chapter 1. The displayed waveform is subject to the inaccuracies caused by the finite bandwidth of the scope, the loading effect, and errors internal to the scope.

5.1 SINE WAVE MEASUREMENTS

Figure 5-1 shows a typical display of a sine wave using an oscilloscope. The usual sine wave parameters can be determined from the display with a small amount of care. The peak-to-peak voltage can be first found in terms of display divisions and then converted to volts. The peak-to-peak value of the sine wave in Figure 5-1 is four divisions. If the vertical sensitivity is set to 0.5 volts per division, then the peak-to-peak voltage is $4 \times 0.5 = 2$ volts. (This value might need to be adjusted if a divider probe is used. See Chapter 4.) The zero-to-peak voltage is just two divisions, so $2 \times 0.5 = 1$ volt.

The RMS voltage is not as easy to determine, at least not directly from the oscilloscope. But the relationship between the zero-to-peak and RMS values for a sine wave is known. $V_{RMS} = 0.707\, V_{0-P} = (0.707)\,(1 \text{ volt}) = 0.707$ volts RMS. This calculation is valid only for a sine wave, but conversion factors for other waveforms were included in Chapter 1. To be precise, the oscilloscope cannot measure RMS

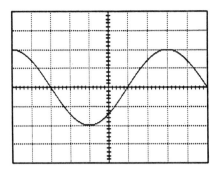

Figure 5-1 A sine wave voltage measured using an oscilloscope. If the vertical sensitivity = 0.5 volts/division, then V_{P-P} = 2 volts and V_{0-P} = 1 volt.

voltage directly, but it does give the user enough information to compute the value for simple waveforms.

The foregoing measurements used the vertical scale to determine voltage information. The horizontal (or time) scale can be used to determine the period of the waveform. The period of the waveform in Figure 5-1 is eight divisions. With the horizontal axis set at 200 μsec/div, the period of the signal is (8 × 200 μsec) = 1.6 msec. Although the frequency cannot be read from the oscilloscope directly, it can be computed using the period of the waveform

$$f = \frac{1}{T}$$

The frequency of the waveform in the figure equals 1/1.6 msec or 625 Hz. Frequency is another parameter which cannot be determined directly from the oscilloscope, but the scope gives us the information to compute the value.

Example 5-1.

Determine the zero-to-peak voltage, the peak-to-peak voltage, period, and frequency of the waveform shown in Figure 5-2.

The waveform is 1.5 divisions zero-to-peak and three divisions peak-to-peak.

V_{0-P} = 1.5 div × 0.2 volts/div = 0.3 volts

V_{P-P} = 3 div × 0.2 volts/div = 0.6 volts

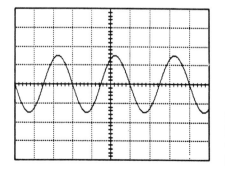

Figure 5-2 Oscilloscope waveform for Example 5-1. Vertical sensitivity = 0.2 volts/division and timebase = 500 μsec/division.

The period of the waveform is three divisions.

$T = 3 \text{ div} \times 500 \text{ } \mu\text{sec/div} = 1.5 \text{ msec}$

$f = \dfrac{1}{T} = 666.67 \text{ Hz}$

5.2 OSCILLOSCOPE VERSUS VOLTMETER

Figure 5-3 shows a voltmeter and an oscilloscope connected to the output of a function generator. A comparison of the two measurements will highlight the advantages of each. The oscilloscope provides a complete representation of the waveform out of the function generator. Its peak-to-peak and zero-to-peak AC values are easily read, and for most waveforms the RMS value can be computed. If there is some DC present along with the waveform, then this, too, will be evident in the voltage versus time display (assuming that the scope is DC coupled). The period and frequency of the waveform can also be measured.

The voltmeter supplies only voltage information about the waveform. Most voltmeters read RMS, but the accuracy of the reading may be dependent on the type of waveform (if the meter is an average-responding type). Also no indication is given as to the actual shape of the waveform. The user may assume a given shape, but distortion due to improper circuit operation may cause the waveform to be much different. The voltmeter is easier to use. It requires less interpretation of its display and has fewer controls to adjust, and its physical size is usually much smaller and more convenient than the scope. Voltmeters are also generally more accurate for amplitude measurements than oscilloscopes. A typical oscilloscope amplitude accuracy is around 3 percent, while voltmeter accuracies are typically better than 1 percent. The voltmeter provides no time or frequency information at all.

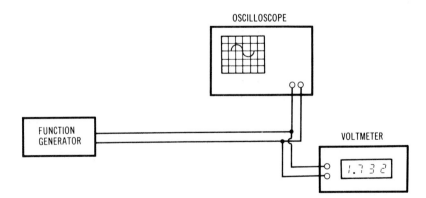

Figure 5-3 The output of a function generator is measured by an oscilloscope and a voltmeter.

5.3 VOLTAGE GAIN MEASUREMENT

One common measurement technique is to apply a sine wave to the input of a circuit and measure the resulting sine wave at the circuit's output (Figure 5-4). The circuit may increase or decrease the waveform amplitude and may also introduce a phase shift. The amplitudes and phases of the two sine waves are compared, giving the voltage gain and phase shift of the circuit.

The input waveform is

$$V_{IN}(t) = V_{IN} \sin(2\pi f t)$$

and the output is

$$V_{OUT}(t) = V_{OUT} \sin(2\pi f t + \theta)$$

Note that the input waveform is defined such that it has a phase shift of zero.

It is often desirable to measure the GAIN of circuits such as amplifiers, filters, and attenuators. The voltage gain is defined as

$$\text{voltage gain} = G = \frac{V_{OUT}}{V_{IN}}$$

where V_{IN} and V_{OUT} are the voltages at the input and output of the circuit, respectively. In other words, gain describes how large the output is compared to the input. V_{IN} and V_{OUT} can be any type of voltage—DC, AC zero-to-peak, AC RMS, and so on—as long as both voltages are measured consistently. If the output is larger than the input, the gain is greater than one; if the output equals the input, then the gain is exactly one; and if the output is less than the input, the gain is less than one. Circuits with gain less than one can be described as having a *loss*.

$$\text{voltage loss} = \frac{V_{IN}}{V_{OUT}} = \frac{1}{G}$$

5.3.1 Gain in Decibels

Voltage gain can also be expressed in dB:

$$\text{voltage gain (dB)} = G_{dB} = 20 \log\left(\frac{V_{OUT}}{V_{IN}}\right)$$

If the output is greater than the input, the gain (in dB) is a positive number. If the output equals the input, the gain is 0 dB, and if the output is less than the input,

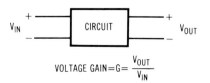

VOLTAGE GAIN $=G= \dfrac{V_{OUT}}{V_{IN}}$

Figure 5-4 The voltage gain of a circuit is determined by dividing the output voltage by the input voltage.

the gain is negative (in dB). Gain and loss are opposite terms. A gain of -10 dB (output is actually smaller than the input) corresponds to a 10-dB loss.

5.3.2 AC voltage gain

Figure 5-5 shows the simplest method for measuring AC voltage gain. A signal source is used to supply the input voltage, and a two-channel oscilloscope is used to measure the input and output voltage of the circuit. The source can be any type capable of producing a sine wave at the frequency of interest. The resulting oscilloscope display is shown in Figure 5-6. The values of the two voltages are determined from the display, and the gain is calculated. If the signal level into the circuit is not critical, then it is desirable to set the sine wave source such that the value of V_{IN} is convenient (such as 1 volt) for the gain calculation. Again, the AC voltage can be described in a variety of ways, but zero-to-peak and peak-to-peak are usually the most convenient.

    It is important to maintain proper loading (both input and output) when making voltage gain measurements. Circuits which expect to be loaded with a particular impedance (Z_0 systems, for instance) should be loaded either with a resistor or an instrument with the appropriate input impedance. Also, the input to such a circuit should be driven with a source that has the correct output impedance.

Example 5-2.

    Determine the voltage gain for V_{IN} and V_{OUT} shown in Figure 5-6. Express the value in dB.

    Figure 5-6 shows the zero-to-peak value of

$$V_{IN} = 1 \text{ div} \times 1 \text{ volt/div} = 1 \text{ volt zero-to-peak}$$

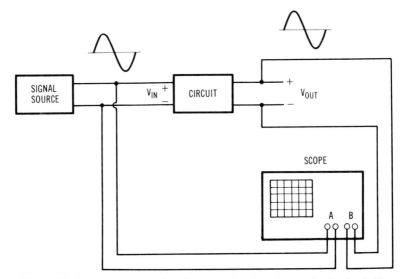

Figure 5-5 Instrument connections for measuring the voltage gain of a circuit. The oscilloscope measures both input and output voltages.

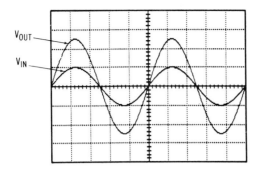

Figure 5-6 The oscilloscope display for measuring the gain of a circuit. Vertical sensitivity = 1 volt/division.

and

V_{OUT} = 2.5 div × 1 volt/div = 2.5 volts zero-to-peak

The voltage gain is

$$G = \frac{V_{OUT}}{V_{IN}} = \frac{2.5}{1} = 2.5$$

Notice how easy the gain calculation was with V_{IN} equal to 1 volt.
In dB,

$$G(\text{dB}) = 20 \log 2.5 = 7.96 \text{ dB}$$

5.4 PHASE MEASUREMENT—TIMEBASE METHOD

The waveforms shown in Figure 5-6 have no phase difference between them, but this is not always the case. Many circuits introduce a phase shift between input and output. In some electronic systems, phase shift is unimportant, but many times it is a critical parameter to be measured.

The same setup shown in Figure 5-5 can be used to measure the phase shift through a circuit. If there is a nonzero phase shift through the circuit, the resulting oscilloscope display will look something like Figure 5-7. (The gain of the circuit is shown as 1 for simplicity.) The scope display gives the user a direct, side-by-side comparison of the two signals, and the phase difference can be determined. First, the

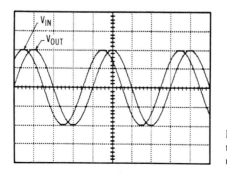

Figure 5-7 The phase difference between two sine waves can be measured using the timebase method.

period of the sine wave is found in terms of graticule divisions. (Recall that one complete cycle corresponds to 360 degrees.) Next, the phase difference is determined in terms of graticule divisions. This can be done best by choosing a convenient spot on one waveform and counting the divisions to the same spot on the other waveform. The starting edge of the sine wave (where it crosses zero) is usually a good reference point, since most scope graticules have the middle of the display marked with the finest resolution. The resulting phase difference is

$$\text{phase shift } \theta = \frac{360 \times \text{shift (in div)}}{\text{period (in div)}}$$

The result is in degrees. Since interpreting the oscilloscope display is somewhat tedious, it is recommended that the user double check the measured value.[1] This simply means take another look at the display, roughly estimate the phase shift (knowing that one cycle is 360 degrees), and compare this to the calculated answer. The two should be roughly the same.

The only problem remaining is to determine the sign (positive or negative) of the phase. V_{OUT} is normally measured with respect to V_{IN}, so V_{IN} is the phase reference. If V_{OUT} is shifted to the left of V_{IN} (V_{OUT} leads V_{IN}), then V_{OUT} will have a positive phase relative to V_{IN}. If V_{OUT} is shifted to the right of V_{IN} (V_{OUT} lags V_{IN}), then V_{OUT} will have a negative phase relative to V_{IN}. The use of the terms lead and lag are less likely to be confused than calling the phase positive or negative.

Since phase repeats on every cycle (360 degrees), the same phase relationship can be described in numerous ways. For example, if V_{OUT} leads V_{IN} by 270 degrees, this will be the same as V_{OUT} lagging V_{IN} by 90 degrees. Although both these expressions are technically correct, it is recommended that phase differences be limited to +180 degrees. So the appropriate expression would be that V_{OUT} lags V_{IN} by 90 degrees.

Example 5-3.

Determine the phase difference between the V_{OUT} and V_{IN} as shown in Figure 5-7.

The period of both waveforms is four divisions. The phase shift in divisions is one-half of a division.

$$\theta = \frac{360 \times 0.5}{4} = 45 \text{ degrees}$$

Since V_{OUT} is shifted to the right of V_{IN}, V_{OUT} lags V_{IN} by 45 degrees, or equivalently, V_{IN} leads V_{OUT} by 45 degrees.

5.5 PHASE MEASUREMENT—LISSAJOUS METHOD

Another method for measuring the phase between two signals is called the LISSA-JOUS METHOD (or Lissajous pattern). Although somewhat more complicated, this method will usually result in a more accurate phase measurement. Figure 5-8

---

[1]This is known as a reality check.

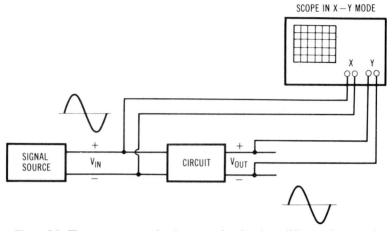

Figure 5-8 The proper connection for measuring the phase difference between the input and output of a circuit using the Lissajous method.

shows a scope connected so that the phase between the output and input of a circuit can be measured. The oscilloscope is configured in the X–Y mode with one signal connected to the horizontal input and the other signal connected to the vertical input.

Figure 5-9 shows the elliptical shape that results from this measurement. Two values, A and B, can be determined from the display and can then be used to calculate the phase angle. The value A is the distance from the X axis to the point where the ellipse crosses the Y axis, and the value B is the height of the ellipse, also measured from the X axis. The scope must be set up with both the X and Y axes set at the zero-volt level. On most scopes this would be accomplished by grounding both inputs and adjusting the dot on the display to be at the center of the screen. Both the volts/division controls can be adjusted to allow convenient and accurate reading of the values. The two controls do *not* have to be set the same, since A and B are measured along the same axis.

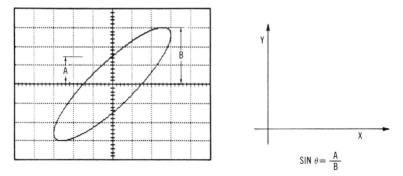

$$SIN\ \theta = \frac{A}{B}$$

Figure 5-9 The oscilloscope is operated in X–Y mode when using the Lissajous method. The two measured values are *A* and *B*, as shown in the figure.

Two general cases must be considered. If the ellipse runs from lower left to upper right, then the phase angle is between 0 and 90 degrees (Figure 5-10). If the ellipse runs from lower right to upper left, then the angle is between 90 and 180 degrees. The angle can be computed from the A and B values using the appropriate equation as shown in Figure 5-10. Unfortunately, the sign of the angle cannot be determined using this method. If the computed answer is 45 degrees, for example, the phase difference may be +45 degrees or −45 degrees. Said another way, the signal on the vertical axis may be leading or lagging the signal on the horizontal axis by 45 degrees. The timebase method can be used to determine the sign, while using the Lissajous method for greater accuracy. A quick look using the timebase method is also a good check on the results from the Lissajous method.

Some special cases of the Lissajous display are shown in Figure 5-11. When the ellipse collapses into a straight line, the two waveforms are in phase. This can be used as a very precise indication when adjusting for zero phase between two signals. If the ellipse is a perfect circle, then the waveforms are exactly 90 degrees apart. Again, this could be plus or minus 90 degrees. If the display becomes a straight line, but in the lower-right–upper-left orientation, then the two signals are exactly out of phase (180 degrees).

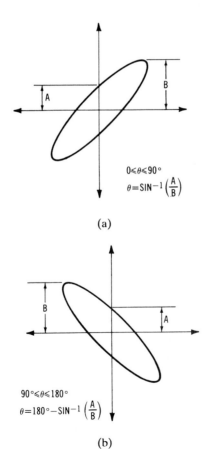

$0 \leqslant \theta \leqslant 90°$

$\theta = SIN^{-1}\left(\dfrac{A}{B}\right)$

(a)

$90° \leqslant \theta \leqslant 180°$

$\theta = 180° - SIN^{-1}\left(\dfrac{A}{B}\right)$

(b)

Figure 5-10 The Lissajous method for computing phase has two general cases. (a) The ellipse runs from lower left to upper right. (b) The ellipse runs from upper left to lower right.

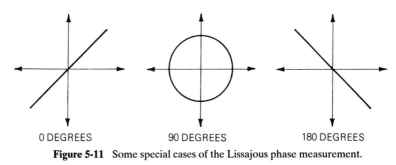

O DEGREES   90 DEGREES   180 DEGREES

Figure 5-11 Some special cases of the Lissajous phase measurement.

Example 5-4.

Determine the phase difference between the two signals given the Lissajous pattern shown in Figure 5-12.

The ellipse is lower left to upper right. First find the values for A and B.

$A = 2.3$ divisions

$B = 3$ divisions

$$\theta = \sin^{-1}\left(\frac{A}{B}\right) = \sin^{-1}\left(\frac{2.3}{3}\right) = 50 \text{ degrees}$$

Note that the vertical sensitivity doesn't enter into the calculation since A and B are measured on the same axis.

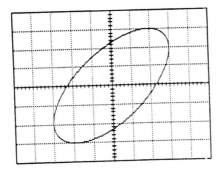

Figure 5-12 Lissajous pattern for Example 5-4.

5.6 FREQUENCY MEASUREMENT—LISSAJOUS METHOD

The Lissajous method can also be used to compare the frequency of two sine waves. The oscilloscope, operating in X–Y mode, is connected as shown in Figure 5-13. The frequency being measured is connected to the vertical axis, while the reference frequency (precisely known, it is hoped) is connected to the horizontal axis. If the two frequencies are the same (have a 1:1 ratio), then the situation is exactly the same as the phase measurement case. In Figure 5-14a, the oscilloscope display is shown for a frequency ratio of 1:1 and a phase shift of 90 degrees. If the phase is other than 90 degrees, then the display will not be a perfect circle, but an ellipse. Again, this case

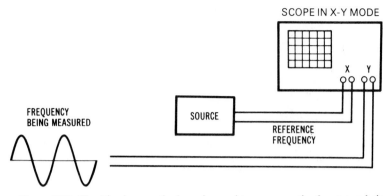

Figure 5-13 The Lissajous method can be used to compare the frequency being measured with a known reference frequency. The oscilloscope is operated in the X–Y mode.

was covered under phase measurement. If the two frequencies are not quite exactly the same, then the display will not be stable and the ellipse will contort and rotate on the display.

Other frequency ratios are also shown in Figure 5-14. They may also appear warped or slanted in different ways just as the ellipse is a slanted version of a circle for the 1:1 case. In general, the ratio of the two frequencies is determined by the number of cusps (or humps) on the top and side of the display. Consider Figure 5-14b. There are two cusps across the top and only one cusp along the side; therefore, the frequency ratio is 2:1. In Figure 5-14d, there are three cusps across the top and two cusps along the side, resulting in a frequency ratio of 3:2. This technique can be applied to any similar frequency ratio. The display will be stable only when the frequency ratio is exact. In general, if the frequency sources are not phase locked together, there will be some residual phase drift between the two frequencies with a corresponding movement on the display.

This method of frequency measurement is somewhat limited since it deals only with distinct frequency ratios. It does help if a sine wave source with variable frequency is available for use as the reference. Then the source can be adjusted so that the measured signal's frequency results in a convenient ratio. Since the method uses frequency ratios, the limiting factor in the accuracy of the measurement is the frequency accuracy and stability of the reference source. If the source used as a frequency reference is not more accurate than the oscilloscope's timebase, then there is no advantage in using this method. Instead, the frequency should be computed from the period of the waveform measured in the timebase mode.

5.7 PULSE MEASUREMENT

Pulse trains (and square waves) can be measured and characterized using an oscilloscope (Figure 5-15). The measurement involves connecting the scope to the waveform of interest, obtaining a voltage versus time display of the waveform and extracting

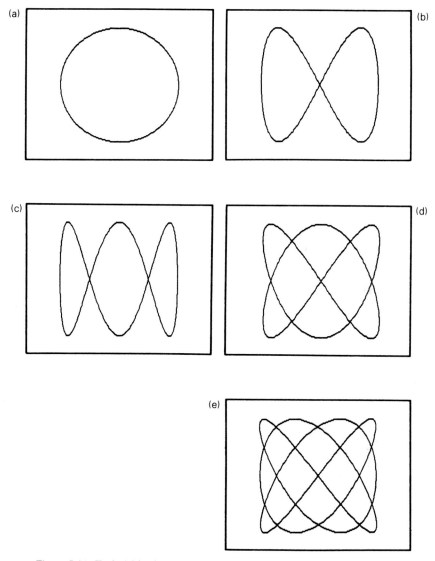

Figure 5-14 Typical Lissajous patterns for a variety of frequency ratios. The frequency ratio is determined by the number of cusps across the top and sides of the pattern. (a) 1:1. (b) 2:1. (c) 3:1. (d) 3:2. (e) 4:3.

the parameter of interest from the display (using the definitions from Chapter 1). The oscilloscope bandwidth and rise time must be adequate so that the pulse being measured is not affected.

Various imperfections may exist in a pulse train (Figure 5-16). Because of the similarities of the pulse train and square wave, these terms for describing imperfections may also be applied to a square wave. Ideally, the pulse goes from 0 volts to V_{0-P} in zero time, but due to circuits that cannot respond infinitely fast, the RISE

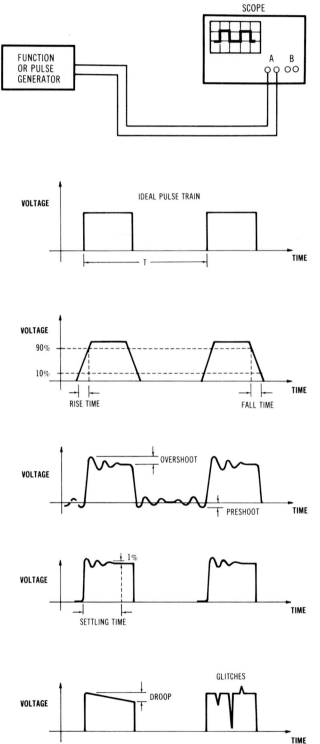

Figure 5-15 The pulse output of a function of pulse generator can be displayed on an oscilloscope.

Figure 5-16 The pulse train introduced in Chapter 1 will usually have some imperfections in it. The quality of the pulse train can be determined using an oscilloscope.

TIME is not zero. Rise time is usually specified to be the time it takes to go from 10 percent of V_{0-P} to 90 percent of V_{0-P}. Similarly, the FALL TIME is the time it takes for the waveform to go from 90 percent of V_{0-P} to 10 percent of V_{0-P}. The rise and fall times may or may not be the same. In some cases, the rise and fall times are specified at the 20 percent and 80 percent levels.

The pulse may actually exceed V_{0-P} after the rising edge of the pulse and then settle out. The amount that the voltage exceeds V_{0-P} is called OVERSHOOT, and the time it takes to settle out is called SETTLING TIME. Settling time is specified so that the waveform has settled to within some small percentage of V_{0-P} (often 1 percent). PRESHOOT is similar to overshoot, except that it occurs before the edge of a pulse.

The top of the pulse may not be perfectly flat, but have some small slope to it. The amount that the top of the pulse slopes down is called DROOP (or SAG). Abrupt voltage changes in the waveform are called GLITCHES and are particularly common in digital circuits. Glitches can be large enough to cause a digital signal to enter the undefined region or in more extreme cases to change logic state. Some circuits will tolerate a certain amount of glitching, but glitches are generally undesirable as they may cause the circuit to malfunction.

5.8 PULSE DELAY

The time delay between two pulses may be measured using a two-channel scope. Figure 5-17 shows two logic gates (inverters) connected end to end driven by a pulsed logic signal. In all digital technologies, it takes a small but nonzero amount of time for the input pulse to reach the output. The oscilloscope is set up to display both the input and output of the two-gate circuit. The resulting time delay between the two waveforms is found by counting the number of divisions between the rising edge of the input pulse and the rising edge of the output pulse. This is then multiplied by the time/division setting on the oscilloscope to obtain the time difference between the pulses.

5.9 DIGITAL SIGNALS

As discussed in Chapter 1, digital signals can take on one of two valid states: HIGH or LOW. Assuming positive logic, these two states may be used to represent the binary numbers 1 and 0, respectively. Thus, most digital signals are pulsed. In general, the pulse width and period will depend on the particular digital system. The digital signal may be periodic, repeating the same binary states in a predictable manner, or it may pulse in what appears to be a random manner, with no identifiable repetition. For instance, a digital signal on the data bus of a microprocessor will change according to the data and instructions being read from memory. Usually, there will be no discernible pattern to these binary numbers, and the digital signal will appear to be a sequence of random pulses. (In addition, the data bus may go into the high-impedance state between transfers, confusing the situation even further.)

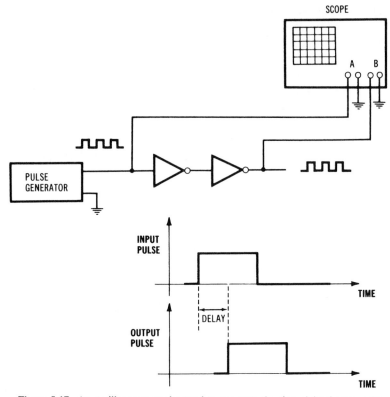

Figure 5-17 An oscilloscope can be used to measure the time delay between two pulses, as in the input and output of a digital circuit.

One should not expect to see perfectly clean pulses in digital signals. Even though digital systems are based on the concept of only two valid states, the actual voltage being observed is still an analog voltage capable of having any value within the power supply range. If the output of the driving circuit goes into the high-impedance state, then the voltage may float to most any value and will be susceptible to cross talk from nearby signals. Also, the digital signal is subject to the same rise time limitations (due to the system bandwidth) as any other signal. Noise can also be present and, if large enough, can cause a digital signal to change state.

Oscilloscopes tend to show all the imperfections of the signal. Sometimes this is unnecessary information that can be ignored but, in many cases, can be critical to understanding the circuit operation or malfunction. A LOGIC ANALYZER is a better choice for viewing digital signals when the analog characteristics of the signal are known to be good (see Chapter 9). Logic analyzers have the advantage of handling large numbers of signals efficiently as well as providing very versatile trace capability. But for critical timing or signal quality measurements, the oscilloscope is the right answer. Some logic analyzers have built-in oscilloscope capability, combining the best of both instruments.

5.9.1 Serial Bit Stream

One common type of digital signal is a serial stream of bits (binary digits). Digital systems, such as the serial interface of a computer, transfer binary numbers one binary digit at a time over a single wire. This can significantly reduce the number of connections required to pass a binary number from one place to another. For instance, an 8-bit number normally requires eight connections (each one representing 1 bit) plus a ground connection. But the same number can be transferred serially using only one signal connection plus ground.

For example, Figure 5-18 shows the voltage waveform for an 8-bit number sent serially. Each bit is transferred within a given time slot. Thus to transfer 8 bits, it takes eight of these time slots. To interpret the waveform correctly, the length of the time slot, the logic polarity (positive or negative logic), and the order of the bits must be known.

If the serial data is sent repetitively, either an analog or digital scope can display the waveform. However, in cases where the serial data is just a single burst, a digital scope's storage capability is valuable for capturing the waveform.

5.9.2 Digital Counter

Other examples of some typical digital waveforms are produced by digital counters. A 4-bit binary counter is shown in Figure 5-19. The binary outputs (Q_0 through Q_3 represent a 4-bit binary number. On each rising edge of the clock input, the binary number increments by one. Thus, the counter counts the number of rising edges of the clock. The binary counting pattern is shown in the figure. Other control lines are often provided to load and clear the counter, but they are not included here.

The voltage waveforms at the digital outputs are plotted in Figure 5-20, assuming that the counter is originally in the 0000 state. Notice that all changes in the outputs only occur on the rising edge of the clock. Its also worth pointing out that the frequency of the Q_0 output is 1/2 the clock frequency (divide by 2). Similarly, the frequency of the Q_1 output is 1/4 the clock frequency (divide by 4), Q_2 is 1/8 the clock frequency (divide by 8), and Q_3 is 1/16 the clock frequency (divide by 16). So a binary

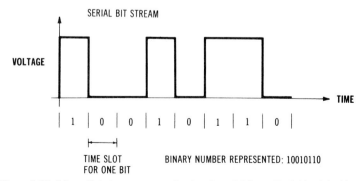

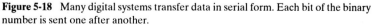

Figure 5-18 Many digital systems transfer data in serial form. Each bit of the binary number is sent one after another.

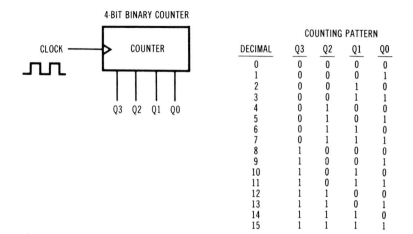

Figure 5-19 A 4-bit binary counter counts the number of rising edges on the clock input. When the count reaches 15, the next rising clock edge will cause the counter to start again at 0.

counter can be used for dividing down a clock to generate submultiple frequencies as well as counting clock pulses. Suppose we need to verify the proper operation of a 4-bit counter. How would we connect an oscilloscope to the counter, and what would be the best signal to trigger on? Figure 5-20 shows the five logic signals associated with the counter. First, we might check the clock signal is present by viewing it with the scope. Then, we should check the timing relationship of Q_0, Q_1, Q_2, and Q_3. One might be tempted to trigger the scope on the rising edge of the clock, since it is the counter's main timing signal. There are 16 rising edges of the clock in Figure 5-20, which means that there are 16 possible trigger events. Sometimes the scope will trigger on the first edge, sometimes on the second edge, and so forth. The oscilloscope display of Q_0 through Q_3 will not be stable as the various trigger events are used. Since the waveforms will move around, measuring the timing relationships will be difficult. Adjusting the trigger holdoff control so that it equals 16 clock periods will stabilize the waveforms, but another solution is to choose a different trigger source. Note that Q_3, the lowest frequency in the system, has only one rising edge in Figure 5-20. Of course, Q_3 will have other trigger events as the signal repeats beyond the edge of the figure, but each time Q_3 goes high, the remaining signals will have exactly the same logic state. Therefore, triggering on the lowest-frequency signal causes the higher-frequency signals to be stable on the oscilloscope display. Once a stable display is obtained, the timing relationships of the signals can be verified. A four-channel scope is desirable here since it can display all four Q outputs simultaneously, while a two-channel scope would force the user to move the scope probes from signal to signal.

The preceding discussion used the example of a 4-bit counter, but the principles can be extended to more complex digital circuits. When viewing multiple signals derived from the same master clock, triggering on the lowest-frequency signal will often be a good choice since it usually makes the other signals stable. Also, trigger holdoff can be used to stabilize troublesome signals.

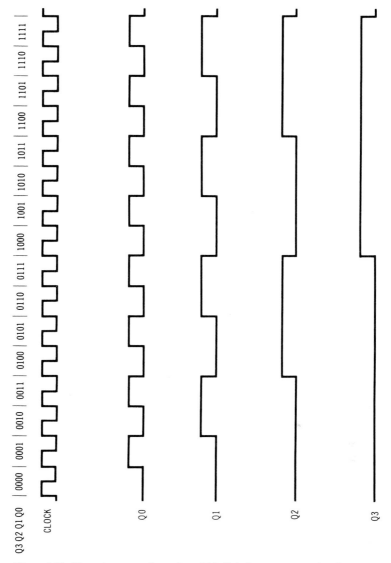

Figure 5-20 The voltage waveforms for a 4-bit digital counter, assuming the counter is initially in the zero state.

5.9.3 D-Type Flip-flop

A D-type flip-flop has an input (D), an output (Q), and a clock input (Figure 5-21). The output remains in the same state until a rising edge of the clock is encountered. At that time, Q takes on the logic level present at the D input. Q will remain in the state until the next rising edge of the clock. The net result is that the Q output tends to track the D input, but only changes on a rising clock edge.

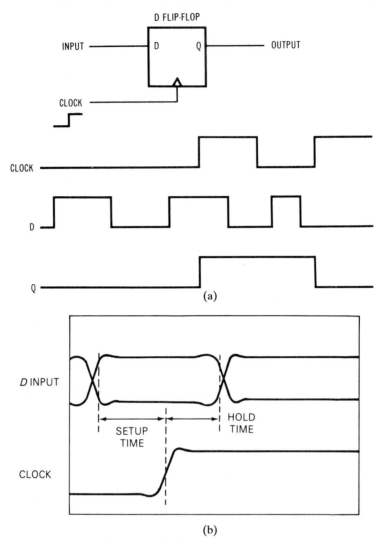

Figure 5-21 (a) The Q output of a D flip-flop changes to the same logic level as the D input whenever a rising edge of the clock occurs. (b) The setup time and hold time for the D flip-flop can be measured using a scope.

Some typical voltage waveforms are shown in Figure 5-21a. The key to understanding the circuit operation and the waveforms is to focus in on the rising clock edges. Only then can the output change state. On the first rising edge of the clock, Q changes from 0 (its previous state) to 1 (the logic level currently present at the D input). Q stays in this state until the second rising clock edge at which time D is 0, so Q becomes 0. Notice that although the logic level at the D input changes several times, these changes are transferred to the Q output only upon a rising edge of the clock. Also, no action occurs on the falling edge of the clock.

For the flip-flop to operate correctly, the D input must be stable at a valid logic level for some time before the rising clock edge. This time is called the SETUP TIME. There may also be a requirement that the D input remains stable for a short time after the clock edge. This time is called the HOLD TIME. On many flip-flops, there is no hold time requirement, which is to say that the required hold time is zero.

A scope can be used to check for setup and hold time violations by viewing the clock and D input and Q output signals while triggering on the rising clock edge (Figure 5-21b). To meet the setup time requirement, the D input must be above the high-logic threshold or below the low-logic threshold during the setup time period in front of the clock edge. The hold time requirement means that the D input must remain a valid logic high or low after the rising clock edge until the hold time expires.

A digital scope with infinite persistence capability is quite useful for checking setup and hold times. In this mode, the scope will capture and retain many occurrences of the D input waveform so that the worse case timing situation can be determined.

It should be apparent from the foregoing examples that digital circuit operation can produce some very complex waveforms. Although these waveforms are digital, we may still need to examine their voltage and timing characteristics. Clearly, a multiple-channel oscilloscope would be of great value in measuring digital circuits. Adjusting the trigger for a stable display is often challenging since the waveforms may not be consistent from sweep to sweep. Try experimenting with the trigger level and slope as well as changing the trigger source. Again, trigger holdoff will often help to stabilize complex waveforms.

One way to sidestep the triggering problem is to capture the signals using a high-sample-rate digital scope set for single sweep. On the first trigger, the scope will capture the waveforms and freeze them in time, allowing the scope user to examine the waveforms. Subsequent triggers are ignored until the scope is reset for another acquisition.

5.10 FREQUENCY RESPONSE MEASUREMENT

Earlier in the chapter, gain and phase measurements were discussed as applied to a single frequency. A single-frequency sine wave was connected to the input of a circuit, and the gain through the circuit as well as the phase of the output signal (relative to the input) were measured. This describes the behavior of that circuit at that particular frequency, but it is often desirable to characterize the circuit performance over a wide range of frequencies. The gain, and to a lesser extent phase, measured at a range of frequencies is called the FREQUENCY RESPONSE of the circuit.

5.10.1 Frequency Response from Single-Frequency Measurements

The most obvious way to measure the frequency response of a circuit is to perform multiple single-frequency gain and phase measurements and plot them as gain versus frequency and phase versus frequency. For example, Figure 5-22 shows a simple RC low-pass filter being driven by a sine wave source. The oscilloscope is connected such that it measures the input voltage and output voltage of the circuit. The resulting voltage measurements at the input and output for a variety of frequencies are tabulated in Table 5-1. Note that if only the gain is required, then other instruments such as a voltmeter could be used to measure V_{IN} and V_{OUT}. One might be tempted

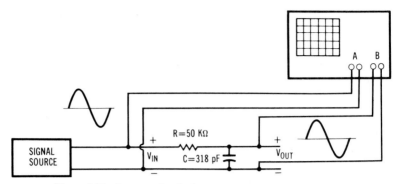

Figure 5-22 An example of a frequency response measurement.

to assume that V_{IN} will always be constant, but this will not necessarily be true. V_{IN} can change due to source flatness and/or loading effects changing with frequency.

The gain for each frequency is calculated by dividing the output voltage by the input voltage at that frequency. Then the frequency response can be plotted as shown in Figure 5-23. Alternatively, the frequency response can be plotted in decibels on the vertical axis and logarithmic frequency on the horizontal axis. Since the dB scale is inherently logarithmic, this results in a log versus log type of plot. This logarithmic scale has the effect of showing widely varying gain values on a compact plot. Notice that the frequency scale of Figure 5-24 easily accommodates two decades of frequency, while the linear scale used in Figure 5-23 does not.

TABLE 5-1 MEASURED VALUES USED TO DETERMINE THE FREQUENCY RESPONSE OF THE RC FILTER.

	Measured Values for RC Circuit			
Frequency	V_{IN}	V_{OUT}	Gain	Gain (dB)
2,000	0.2	0.20	0.98	−0.17
4,000	0.2	0.19	0.93	−0.64
6,000	0.2	0.17	0.86	−1.33
8,000	0.2	0.16	0.78	−2.15
10,000	0.2	0.14	0.71	−3.01
12,000	0.2	0.13	0.64	−3.87
14,000	0.2	0.12	0.58	−4.71
16,000	0.2	0.11	0.53	−5.51
18,000	0.2	0.10	0.49	−6.27
20,000	0.2	0.09	0.45	−6.98
30,000	0.2	0.06	0.32	−9.99
40,000	0.2	0.05	0.24	−12.30
50,000	0.2	0.04	0.20	−14.14
60,000	0.2	0.03	0.16	−15.67
70,000	0.2	0.03	0.14	−16.98
80,000	0.2	0.02	0.12	−18.12
90,000	0.2	0.02	0.11	−19.13
100,000	0.2	0.02	0.10	−20.03

5.10.2 Swept-frequency Response

Although the previously described point-by-point method is valid and produces accurate results, it is somewhat time consuming. Another method of measuring frequency response involves using a sweep generator to speed up the measurement. The swept sine wave of the sweep generator is connected to the input of the circuit under test (Figure 5-25). The output of the circuit under test is connected to the

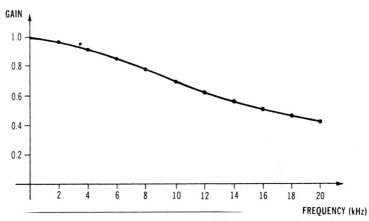

Figure 5-23 The frequency response of the RC filter plotted as linear gain versus linear frequency.

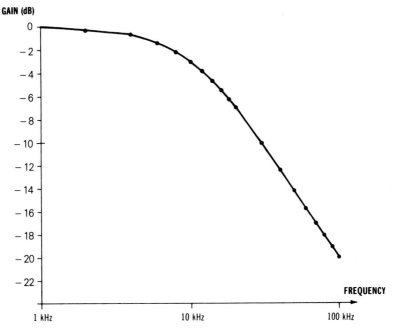

Figure 5-24 The frequency response of the RC filter plotted as gain in decibels versus log frequency. Note how a wider range of both gain and frequency can be represented than with the linear plot.

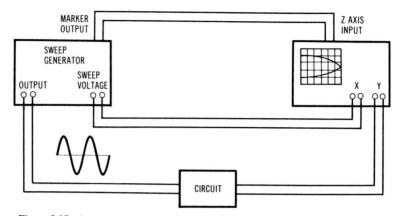

Figure 5-25 A sweep generator and a scope operating in X–Y mode can be used to plot the frequency response of the circuit automatically.

vertical channel of an oscilloscope, operating in the X–Y mode. The sweep voltage of the sweep generator drives the horizontal axis of the scope. As the sweep generator sweeps in frequency, the sweep voltage of the generator ramps up (in proportion to frequency), causing the output of the circuit under test to be plotted across the scope display. In this manner, the entire frequency response of the circuit is quickly displayed on the scope (Figure 5-26).

This method relies on the output of the sweep generator being constant with frequency. The flatness of the generator will cause an error in the frequency response. Also the drive capability of the generator is important. The load that the circuit places on the generator will usually vary over frequency. The generator must be relatively insensitive to these load changes, or another error will be introduced. For both these reasons, it is a good idea to check that V_{IN} is constant as the generator sweeps.

The oscilloscope is displaying V_{OUT}, and not the gain (V_{OUT}/V_{IN}). If V_{IN} was set up to be a convenient value (such as one volt) then the display can be interpreted directly as gain. Otherwise, a small amount of mental arithmetic may be necessary to convert the V_{OUT} display to actual gain.

The sweep generator should not be swept too fast, since the circuit under test needs time to respond. This is particularly important in circuits with abrupt changes

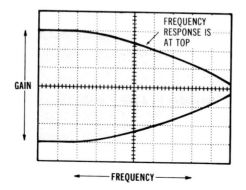

Figure 5-26 The oscilloscope display of the swept-frequency response. The vertical axis is V_{OUT}, and the horizontal axis is the frequency range swept by the generator.

in gain as the frequency varies. The sweep rate for the generator is usually set experimentally by reducing the sweep rate until the frequency response no longer changes with each change in sweep rate. The sweep generator may be swept in either a linear or logarithmic manner depending on the desired type of frequency axis. The oscilloscope vertical axis is, of course, always linear.

It is sometimes desirable to locate a particular point on the frequency response curve very precisely with respect to frequency. Thus, many sweep generators supply a MARKER output signal that pulses when a particular frequency (or frequencies) is present at the generator output. This signal can be connected to the Z-axis input of the oscilloscope. The sweep generator will pulse the marker output, causing a change in intensity on the oscilloscope display at precisely the marker frequency. Exactly how the intensity changes (whether it gets brighter or dimmer) will depend on the polarity of the marker signal as well as the polarity of the Z-axis input. Another type of marker function is sometimes provided which pulses the output voltage slightly at the marker frequency. This causes a "blip" to appear on the display at the marker frequency. The Z-axis input is not used in this case.

5.11 SQUARE WAVE TESTING

Sine waves are the predominant signal for characterizing the frequency response of analog circuits, but the square wave can also be used. This technique is valid only for devices with flat frequency responses, such as audio amplifiers. (A standard test for audio equipment is a 1-kHz square wave test.) In addition, the circuit under test must be DC coupled or AC coupled to well below the square wave's frequency. One special case of square wave testing is the compensation of attenuating probes already discussed in Chapter 4.

A square wave is applied to the input of the circuit being tested, and the output of the circuit is monitored on an oscilloscope (Figure 5-27). The square wave, as discussed in Chapter 1, is very rich in harmonics, extending out to many times its fundamental frequency. The relative amplitude of each of these harmonics must remain unchanged for the output to be a square wave. If the circuit under test is an amplifier, the amplitude of each harmonic is increased by the amplifier's gain, but

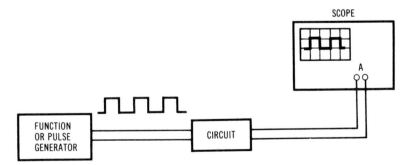

Figure 5-27   The square wave test of a circuit is a good qualitative test of a circuit's frequency response.

their amplitude relative to each other should remain the same. In addition, each harmonic has a particular phase relationship with the fundamental frequency that must be maintained. Otherwise, the output will not be a true square wave. The circuit could even have a perfectly flat amplitude response, but not pass a square wave correctly due to phase distortion. This type of test requires a high-quality waveform at the input; otherwise, the output square wave will also be degraded. Depending on the output characteristics of the generator, the circuit under test may load the generator enough to cause distortion. It is usually a good idea to monitor the input waveform with the second channel of the oscilloscope.

Although the square wave test does not result in a frequency response plot, it does test a circuit quickly, with good qualitative results. Some typical output waveforms encountered in square wave testing and their causes are shown in Fig. 5-28. Phase shift at low or high frequencies can cause a tilt to one side or the other of the square wave. It is difficult to predict the effects of attenuation at either high or low frequencies as it depends greatly on the exact shape of the frequency response. However, a few typical examples are shown. Some circuits will exhibit a noticeable degradation in the rise time of the output square wave. This is referred to as slew rate limiting. Due to the nature of the square wave test, it is more effective at determining whether a problem exists than identifying the particular problem.

LOW-FREQUENCY AMPLITUDE LOSS

HIGH-FREQUENCY AMPLITUDE LOSS

LOW-FREQUENCY AMPLITUDE LOSS
OR AC COUPLING

HIGH-FREQUENCY AMPLITUDE LOSS

RESONANCE OR HIGH-FREQUENCY LOSS

SLEW-RATE LIMITING

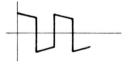

HIGH-FREQUENCY PHASE LAGGING

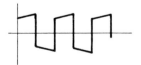

LOW-FREQUENCY PHASE LAGGING

Figure 5-28   Some examples of waveforms resulting from the square wave test.

5.12 LINEARITY MEASUREMENT

It is often desirable to measure the DC voltage out of a circuit compared to the DC voltage in. This can be done using the X–Y mode of the oscilloscope as shown in Fig. 5-29. A slowly varying DC voltage is applied to the input of the circuit as well as the X axis of the scope. A convenient method of obtaining the varying DC voltage is to use a triangle wave with a very low frequency. The output of the circuit is connected to the Y axis of the scope, resulting in the output voltage being plotted versus the input voltage. The frequency of the generator is chosen fast enough so that the display does not flicker too much, but slow enough that the operation of the circuit is not affected. (Remember, a DC voltage is being simulated.)

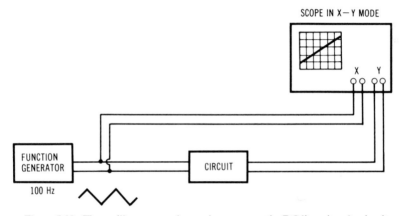

Figure 5-29 The oscilloscope can be used to measure the DC linearity of a circuit.

Consider the clipping circuit shown in Fig. 5-30. For V_{IN} less than 5 volts and greater than 0 volts, the Zener diode acts like an open circuit and V_{OUT} is equal to V_{IN}. When V_{IN} becomes greater than 5 volts (the Zener voltage), then the diode turns on and V_{OUT} is limited to 5 volts, no matter how large V_{IN} gets. If V_{IN} becomes less than zero, then the diode turns on in the other direction and limits V_{OUT} to 0 volts. (Actually, it would limit the voltage at a slightly negative value, typically −0.6 volts, depending on the diode.) At any rate, the effect of the circuit is to limit the output voltage to between about 0 and 5 volts.

If a DC linearity measurement were made on the circuit using the technique described, then the display would appear as shown in Fig. 5-31. The sloped portion of

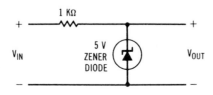

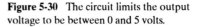

Figure 5-30 The circuit limits the output voltage to be between 0 and 5 volts.

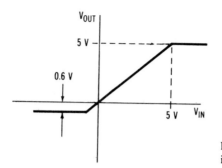

Figure 5-31 The V_{OUT} versus V_{IN} linearity display for the clipping circuit.

the trace corresponds to the region where V_{OUT} equals V_{IN}. The two flat parts of the trace are where V_{OUT} is limited by the clipping action of the circuit.

    With an amplifier, it is usually desirable to have the output be an exact replica of the input, but amplified by some amount. This is expressed mathematically as

$$V_{OUT} = kV_{IN}$$

where k is the voltage gain of the amplifier.

    Thus, the output waveform is the same as the input waveform, except with a larger amplitude. For increasingly larger input voltages, the amplifier will at some point stop producing a proportionally larger output voltage. At this point, the output waveform will be clipped (Fig. 5-32). The peaks of the waveform are flattened out at this output level.

    The V_{OUT} versus V_{IN} linearity plot of the amplifier can also be used to measure this phenomenon (Fig. 5-33a). The display is a straight line whose slope is the gain of the amplifier. Ideally, the straight line of the display would extend indefinitely. That is, the amplifier would be capable of amplifying any input voltage, no matter how large. In reality, the amplifier will clip at some point, usually as the peak voltage approaches the amplifier's power supply voltage, as shown in Figure 5-33b. The

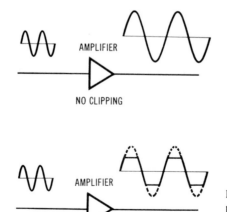

Figure 5-32 Clipping occurs at the output of an amplifier when the amplifier cannot produce the peak voltage of the waveform.

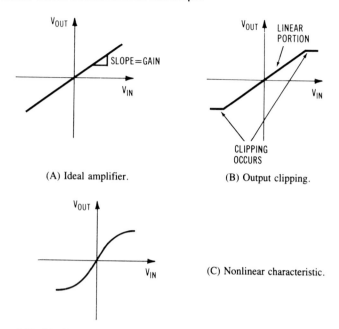

(A) Ideal amplifier.   (B) Output clipping.

(C) Nonlinear characteristic.

Figure 5-33   The V_{OUT} versus V_{IN} plot for an ideal amplifier is a straight line. In a real amplifier, imperfections such as output clipping and other nonlinearities may occur.

straight line is still present, but flattens out at the point of limiting. Typically, the trace does not break sharply, but instead is rounded off.

Since the V_{OUT} versus V_{IN} characteristic is a straight line, this type of circuit operation is called LINEAR. If the plot is not a straight line, the characteristic is termed nonlinear (Figure 5-33c). Nonlinear amplifier operation causes distortion of the output signal (usually in the form of harmonics).

5.13 CURVE TRACER MEASUREMENT TECHNIQUE

Another useful oscilloscope measurement technique is the curve tracer circuit. This technique uses the oscilloscope in X–Y mode to display the current through a component (such as a resistor or diode) versus the voltage across the component. This display is referred to as the *I–V* (current-voltage) characteristic of the component.

There are several ways to make this measurement. Figure 5-34a shows a function generator driving the curve tracer circuit. V_2 is the voltage across the device being measured, and V_1 is the voltage across the resistor. By Ohm's law, V_1 is proportional to the current through the resistor, which is also the current through the component under test. So if V_1 could be displayed versus V_2, the *I–V* characteristic of the component would be shown. If the scope has floating inputs, then this can be done with no problem.

Unfortunately, most scopes have grounded inputs, which results in the situation shown in Figure 5-34b. Both sides of the component under test are grounded, resulting

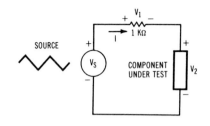

(a)

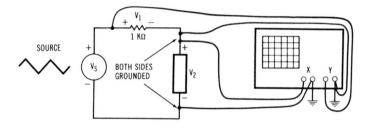

(b)

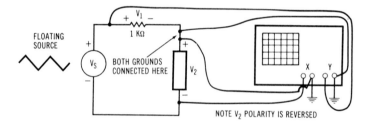

(c)

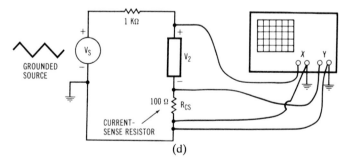

(d)

Figure 5-34 Methods for displaying the *I–V* characteristic of a component. (a) The approach suitable for use with floating scope inputs. (b) Grounding problem with typical scope with grounded inputs. (c) An approach for use with a floating source. (d) Use of a current-sense resistor.

in a short circuit across it. The situation is compounded even further if the source has a grounded output. If the source is floating, then the circuit in Figure 5-34c can be used. Note that the circuit ends up being grounded at only one point, which is where both scope input grounds are connected. Something has changed, though. The horizontal axis of the scope will be $-V_2$ (instead of $+V_2$) due to the reversal of the horizontal input leads. This can be compensated for if the scope has a "invert channel" switch; otherwise, the I–V curve will appear backward on the display (left half of the display swapped with the right half). This may be acceptable if the user is willing to convert it back mentally.

If both the scope and the source are grounded, then another technique must be used. A current-sense resistor is placed in series with the component under test. The vertical scope input then uses the voltage across this resistor to measure the current. The voltage across the current-sense resistor will introduce a small error in the measurement of V_2, but as long as the resistor is kept small, the error will be acceptable. The resistor cannot be made too small since, for a given current being measured, the voltage will decrease with smaller resistance. The sensitivity of the scope will ultimately determine how small the current-sense resistor can be made.

5.13.1 Diode *I–V* Characteristic

Fig. 5-35 shows an oscilloscope set up to measure the I–V characteristics of a diode, using the current-sense method. The source (usually a function generator) is set to produce a low-frequency triangle wave, although a sine wave will also work. The triangle wave acts as an automatically varying DC voltage, causing the voltage across the diode to also change. At the same time, the voltage across and the current through the diode are measured. The amplitude and frequency of the function generator can be set experimentally, but a zero-to-peak voltage of 5 volts and a frequency of 30 Hz is a good starting point.

The resulting I–V characteristic for a diode is shown in Fig. 5-36. The horizontal scale can be determined directly from the volts/division setting. The vertical scale must take into account the value of the current senses resistor.

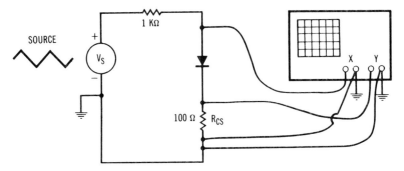

Figure 5-35   The *I–V* characteristic of a diode can be measured using the curve tracer circuit with current-sensing resistor.

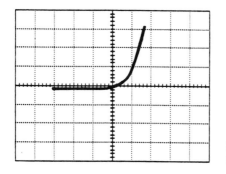

Figure 5-36 The *I–V* curve of a solid-
state diode.

$$\text{amps per division} = \frac{\text{volts per division}}{R}$$

where R is the value of the current sense resistor.

The resulting curve is the classical behavior of the solid-state diode. For voltages greater than zero (right half of the display), the current quickly increases. For voltages less than zero (left half of the display), the current is essentially zero. So the diode conducts in the forward direction, but acts like an open circuit in the reverse direction.

The scope display may actually show two separate traces instead of the single curve. One trace is drawn as the voltage increases, and the other is drawn on the decreasing portion of the triangle wave. They may be slightly different due to either capacitive effects or heating of the component being tested. Reducing either the frequency or the amplitude of the triangle wave will cause the two traces to converge into one single trace.

5.13.2 Resistor *I–V* Characteristic

It is worth noting at this point that the curve tracer circuit can be used to measure unknown resistors. The resulting *I–V* curve is shown in Figure 5-37. The current-sense resistor, R_{CS} should be chosen to be at least a factor of 10 smaller than the unknown resistance. The voltages shown in Fig. 5-37 (ΔX and ΔY) should be determined, taking into account the volts/division settings. The resistance value is determined by calculating 1/slope of the line, using the equation that follows.

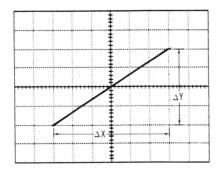

Figure 5-37 The *I–V* curve of a resistor
is a straight line with $R = 1/\text{slope}$.

$$R = \frac{1}{\text{slope}}$$

$$= \frac{\Delta X}{\Delta Y} R_{\text{CS}}$$

5.14 AMPLITUDE MODULATION MEASUREMENT

An amplitude modulated signal is represented by the equation

$$v(t) = A_{\text{C}}\,[1 + am(t)]\cos(2\pi\,f_c t)$$

where

A_{C} = the signal amplitude
a = modulation index $(0 \le a \le 1)$
$m(t)$ = normalized modulating signal (maximum value is 1)
f_c = carrier frequency

The modulation percentage of an amplitude modulated signal can be measured in the time domain with an oscilloscope. The timebase of the scope is set up to view the modulating frequency (not the carrier frequency). As shown in Fig. 5-38, the modulating frequency shows up in the envelope of the carrier. The modulation percentage is determined by noting the maximum and minimum of the envelope, V_{MAX} and V_{MIN}. The maximum of the envelope voltage occurs when the modulating signal is at its most positive value, +1, and the minimum occurs when the modulating signal is −1.

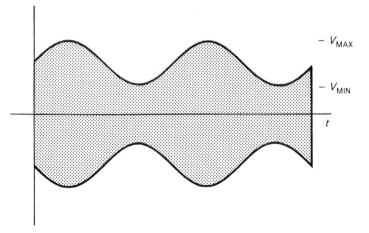

Figure 5-38 An amplitude modulation measurement is made using the minimum and maximum of the envelope of the signal.

$$V_{\text{MAX}} = 1 + a$$
$$V_{\text{MIN}} = 1 - a$$

Solving for the modulation index,

$$a = \frac{V_{\text{max}} - V_{\text{min}}}{V_{\text{max}} + V_{\text{min}}}$$

If the modulating signal is available, it can be used as the trigger source. If not, the modulated signal can be stabilized by triggering near the top of the waveform.

REFERENCES

OLIVER, BERNARD M., and JOHN M. CAGE. *Electronic Measurements and Instrumentation.* New York: McGraw-Hill Book Company, 1971.

WEDLOCK, BRUCE D., and JAMES K. ROBERGE. *Electronic Components and Measurements.* Englewood Cliffs, NJ: Prentice Hall, Inc., 1969.

LENK, JOHN D. *Handbook of Oscilloscopes.* Englewood Cliffs, NJ: Prentice Hall, Inc., 1982.

WITTE, ROBERT A. *Spectrum and Network Measurements.* Englewood Cliffs, NJ: Prentice Hall, Inc., 1991.

chapter six

Frequency Counters

This chapter discusses a specialized measuring instrument for measuring the frequency of a signal—the frequency counter. Frequency counters use stable crystal oscillators and digital counting circuits to provide simple, nonerror-prone frequency and period measurements. Some frequency counters are capable of measuring other parameters such as risetime and pulse width. Frequency counters are commonly used for radio frequency measurements, but they can also be found in use at lower frequencies.

6.1 FREQUENCY COUNTERS

As covered in Chapter 1, the frequency of a periodic waveform is defined as the number of cycles that occur per second. A convenient and accurate way of measuring frequency is using a frequency counter. The measurement is usually as simple as connecting the frequency counter to the signal being measured and reading the digital display.

The frequency of a signal is defined by how often zero crossings occur. To a frequency counter, the definition of frequency is

$$f = \frac{n}{T}$$

where

n = the number of cycles of the waveform
T = the time interval over which the cycles are counted

Figure 6-1 shows a conceptual block diagram of a frequency counter. The signal being measured is amplified and changed into a digital pulse train. This pulse train

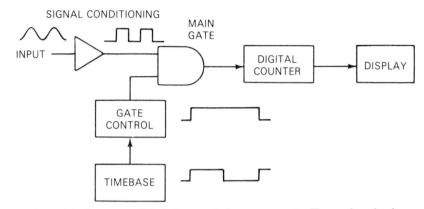

Figure 6-1 A conceptual block diagram of a frequency counter. The number of cycles in the signal being measured is counted for a length of time determined by the timebase.

passes through an electronic switch called the MAIN GATE and drives a series of digital counters. If the main gate is opened, the value of the digital counters increases by one for each new cycle of the signal being measured. If the main gate were to remain opened, the digital counter would keep counting up indefinitely (or at least until it ran out of digits). Instead, the main gate is opened for a known length of time, and the resulting number of cycles of the waveform is measured. This number represents the frequency of the waveform. To perform another measurement, the digital counter is reset, and the main gate is once again reopened. The waveforms associated with this operation are shown in Figure 6-2.

As an example, suppose that the length of time that the main gate is closed is one second. The number of cycles in one second (which is the frequency in hertz) would be displayed on the digital counter. The measurement time of one second was chosen for simplicity in explaining the operation of the frequency counter. For high-frequency signals, the one-second gate time would cause the digital counters to reach its maximum count and overflow. For these signals a shorter gate time is needed (and for slower signals a longer gate time is needed). These different gate times correspond to different measurement ranges of a counter.

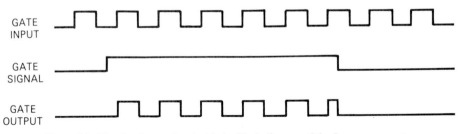

Figure 6-2 The signals associated with the block diagram of the frequency counter. The gate signal turns on the main gate, allowing the input signal to pass through to the digital counter.

In Figure 6-2, note that the last pulse at the end of the gate output is shorter than the rest of the pulses. This is due to the main gate closing midway through the period of the input signal. Had the gate closed slightly earlier, this pulse would have been completely suppressed. Had the gate closed later, the full pulse (and perhaps the start of the next input period) would have been passed on to the digital counter. This illustrates the ±1 count ambiguity which is typical of counter measurements.

A typical frequency counter is shown in Figure 6-3.

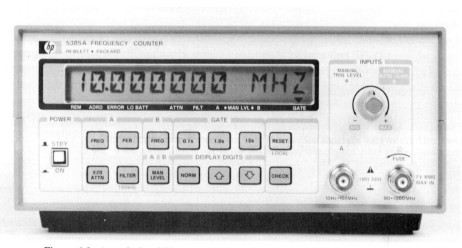

Figure 6-3 A typical 1-GHz frequency counter with both frequency and period measurement functions. (Photo courtesy of Hewlett-Packard Company.)

6.2 FREQUENCY DIVIDERS

A FREQUENCY DIVIDER is used to reduce the frequency of a digital signal. Figure 6-4a shows a frequency divider which divides the input frequency by 2 to produce an output frequency at half the input frequency. Similarly, Figure 6-4b shows a divide-by-10 circuit which reduces its input frequency by a factor of 10. Notice that both the input and output of the divider is a pulse train. This type of circuit can be used to increase the range of a frequency counter in two ways.

Figure 6-5 shows a divide-by-10 circuit added to the frequency counter in two different places. At the input, the frequency divider has the effect of increasing the maximum measurable frequency by reducing incoming signals to a range that is usable by the basic frequency counter. In this mode, frequency dividers are often referred to as PRESCALERS. For instance, the basic frequency counter may have a maximum frequency limitation of 10 MHz. Adding a divide-by-10 circuit in front of the basic counter extends the range by a factor of 10 to 100 MHz. The prescaler also causes the resolution of the measurement to be reduced by a factor of 10.

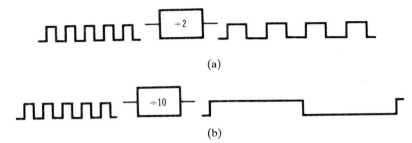

(a)

(b)

Figure 6-4 A frequency divider produces an output frequency which is equal to the input frequency divided by an integer number. (a) A divide-by-2 frequency divider. (b) A divide-by-10 frequency divider.

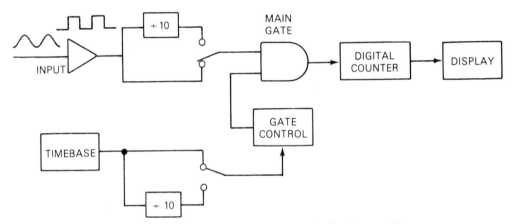

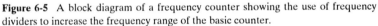

Figure 6-5 A block diagram of a frequency counter showing the use of frequency dividers to increase the frequency range of the basic counter.

The other frequency divider was added to the timebase circuit. This has the effect of increasing the length of time that the main gate is on, which means that lower frequencies can be measured than with the original frequency counter. Another benefit of the increased gate time is improved frequency resolution.

In both cases, the frequency divider is shown as being able to be switched in and out of the circuit. Typically, several divider circuits are supplied so that the user can conveniently select a measurement range on the instrument. Sometimes a special high-frequency prescaler is offered as an external option that increases the high-frequency range of a counter.

6.3 PERIOD MEASUREMENT

The user of a frequency counter can determine the period of a waveform from its frequency $(f = 1/T)$, but period measurement is often built into a frequency counter. Figure 6-6 shows a small, but important, change in the basic frequency

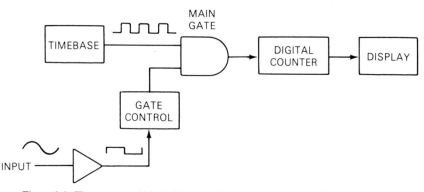

Figure 6-6 The conceptual block diagram of a frequency counter while measuring the period of the input waveform.

counter block diagram—the input and timebase connections are interchanged. In this mode, the input opens the main gate for one of its cycles. During this cycle (which is the period of the input), the number of timebase clocks are counted. Suppose the timebase period was 1 msec; then the resulting display would be the number of 1-msec cycles that occurred during one cycle of the input waveform. Put simply, this is the period of the input waveform in milliseconds. By using selectable frequency dividers, other timebase frequencies are easily generated, resulting in other ranges of period measurement.

6.4 RECIPROCAL COUNTER

The two block diagrams associated with frequency measurement and period measurement have their own advantages and disadvantages. For low frequencies, period measurement has higher resolution and is faster. For frequencies above the timebase clock, the frequency measuring block diagram is better. The reciprocal counter combines the best of both counting techniques, using period mode for low frequencies and frequency mode for high frequencies.

Considering a numerical example will help highlight the advantages and disadvantages of each counting technique. Suppose a counter has a timebase clock of 10 MHz and can operate in either the frequency or period mode. First, consider the case where the input frequency is very low, for example, 1 Hz. In the period mode, the counter can produce the result in one cycle of the input waveform, in this case, 1 sec. The period measurement has an uncertainty of ± 1 count of the 10-MHz timebase, or ± 100 nsec. In frequency mode, the counter counts the number of input cycles which occur during some measurement time. As an example, let's assume that the measurement time is 10 sec (which happens to be much longer than the time the period measurement took). During 10 sec, 10 input cycles occur, and the counter measures this frequency with an uncertainty of ± 1 count, or ± 1 Hz. The result is that even though the frequency mode takes much longer to complete a measurement, the frequency measurement is much less accurate than the period mode.

Now consider the case where the input frequency is much higher than the timebase frequency. For example, say that the input frequency is 100 MHz, 10 times higher than the timebase frequency. If measured with the period mode of the counter, the 10-nsec period is measured with an uncertainty of ±1 timebase count, or ±100 nsec. In frequency mode, the uncertainty is again ±1 count of the input frequency, or ±10 nsec. So for frequencies higher than the timebase frequency, the frequency mode is more accurate.

The reciprocal counter combines the frequency and period modes and is implemented as shown in Figure 6-7. The block diagram contains two counters: the time counter and the event counter. The time counter counts the timebase periods while the event counter counts the cycles of the input frequency. The period of the input signal is given by

$$T = \frac{T_{\text{CLK}} N_{\text{CLK}}}{N_{\text{EVENT}}}$$

where

$$T_{\text{CLK}} = \text{the timebase clock period}$$
$$N_{\text{CLK}} = \text{the clock count}$$
$$N_{\text{EVENT}} = \text{the event count}$$

The reciprocal counter requires some computational circuitry to perform the period calculation (or its reciprocal, if a frequency measurement is desired). With the availability of low-cost microprocessors, this is not a serious problem.

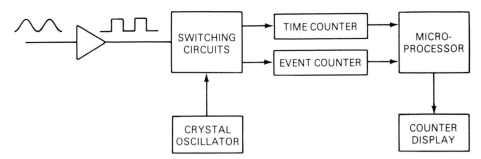

Figure 6-7 The reciprocal counter combines the best features of frequency mode and period mode.

6.5 UNIVERSAL COUNTER

Universal counters extend the basic counter circuitry such that it can measure other waveform parameters. Besides frequency and period, measurements commonly made by universal counters include time interval, pulse width, duty cycle, rise time, fall time, and phase. Time interval measurement and pulse width measurement are a logical extension of a period measurement, with the start and stop times of the measurement no longer limited to the start and stop of one period. Rise time is derived from the time interval circuitry, but with the start and stop times determined by comparators which trigger at the 10 percent and 90 percent points of the waveform (Figure 6-8). Duty cycle is a pulse width measurement divided by the period of the waveform. A phase measurement measures the time delay between two waveforms and expresses in degrees, with 360 degrees corresponding to one period of the waveform.

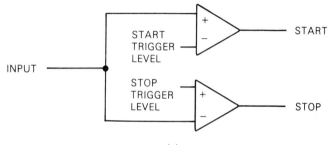

(a)

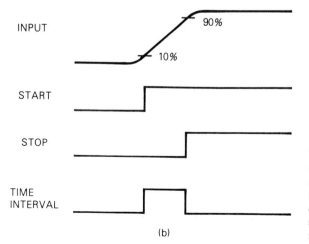

(b)

Figure 6-8 Two comparators are used to convert the rise time of the input signal to a time interval that is measurable by counter circuits. (a) The comparator circuit. (b) The signals associated with the rise time measurement.

6.6 GATED COUNTER MEASUREMENTS

For signals such as pulsed RF waveforms, the sinusoid being measured pulses on and off. Attempts to measure the frequency of such a signal by measuring the number of cycles for a time longer than the pulse width will be in error. Some counters automatically gate the counter circuitry on for a controlled length of time and perform the measurement only while the pulsed RF is active (Figure 6-9).

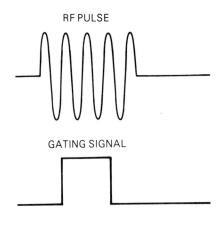

Figure 6-9 A frequency counter with a gated measurement mode can measure the frequency during an RF pulse.

6.7 TIMEBASE ACCURACY

The timebase of a frequency counter is usually a precisely controlled crystal oscillator, which is divided down to produce the needed frequency. This results in a timebase accuracy that is limited only by the stability and accuracy of the crystal oscillator. Since this oscillator must operate at only one frequency, it can be designed to be extremely stable.

There are three main types of crystal oscillators: room temperature crystal oscillators (RTXO), temperature compensated crystal oscillators (TCXO), and oven controlled crystal oscillators (OCXO). Room temperature crystal oscillators are designed to be relatively stable over some temperature range, typically 0° to 50°C. A careful choice of crystal typically results in a temperature stability of 2.5 parts per million (ppm) over this temperature range. A TXCO takes the basic crystal oscillator design and adds a temperature-sensitive component or components which compensate for the inherent temperature characteristics of the crystal. This can produce an order of magnitude improvement in stability, typically 0.5 ppm over the same temperature range. The ultimate solution to frequency stability is to remove the temperature variation that the crystal experiences. An oven controlled crystal oscillator has the crystal housed in an oven which is stabilized in temperature by a heating element. The control system for the heater can be a simple ON/OFF type or a more complex linear control system. Oven oscillators can typically achieve frequency stability of 10 parts in 10^9 or 0.01 ppm over a 0° to 50°C temperature range. Table 6-1 summarizes the stability typically obtained by various types of oscillators.

TABLE 6-1 TYPICAL FREQUENCY STABILITY
FOR OSCILLATORS

Oscillator Type	Frequency Stability
RTXO	2.5 ppm
TCXO	0.5 ppm
Oven oscillator	0.01 ppm

6.8 INPUT IMPEDANCE

Frequency counters generally have either a 50-Ω or a 1-MΩ input impedance. Like several of the instruments already discussed, the 1-MΩ input is convenient for general-purpose low-frequency measurements. The 50-Ω input becomes necessary as higher-frequency measurements are made (typically greater than 50 MHz). Since a frequency counter concerns only frequency (and not amplitude or voltage), a decreased signal level does not immediately reduce measurement accuracy. Thus, capacitive loading due to the 1-MΩ input may reduce the signal level into the frequency counter somewhat without introducing error into the measurement. At some point, however, the signal is attenuated so much that the counter can no longer detect its frequency accurately. Various combinations of 50-Ω and 1-MΩ inputs exist. Some counters include both, with the 1-MΩ input intended for low-frequency operation and the 50-MΩ input for high-frequency operation. Other counters have one input that is switchable between 1 MΩ and 50 Ω.

6.9 FREQUENCY COUNTER SPECIFICATIONS

The specifications for a typical frequency counter are given in Table 6-2. The frequency range is shown as dependent on which input impedance is being used. The sensitivity specification defines how small a signal can be measured, while frequency resolution determines the smallest change in frequency that can be detected.

The most desirable specification is conspicuous by its absence. There is no frequency accuracy spec. This seems a bit strange at first, until one understands what limits the performance of the instrument. If the timebase of a frequency counter is adjusted to be exactly on frequency, presumably by comparing it to some perfect frequency standard, then there is essentially no frequency error at that instant in time. However, over any time interval the timebase tends to drift in frequency, primarily due to aging effects in the crystal oscillator and changing performance with temperature. Fortunately, the aging and temperature stability are usually specified by the manufacturer.

Consider the typical specifications given in Table 6-2. Ignoring temperature changes for the moment, the long-term frequency stability (with the standard time base) is determined by the aging rate (0.1 ppm/month). Assuming the frequency counter was adjusted to be exactly on frequency (to within the resolution of the

TABLE 6-2 TYPICAL FREQUENCY COUNTER SPECIFICATIONS

Specification	Input A	Input B
Frequency range	10 Hz to 100 MHz	90 MHz to 1000 MHz
Input impedance	1 MΩ with	50Ω
	25 pF capacitance	
Sensitivity	15 mV RMS	10 mV RMS
Frequency resolution	8 digits	
Temperature stability		
Standard timebase	<2 ppm, 0° to 40°C	
Optional oven timebase	<0.1 ppm, 0° to 50°C	
Aging rate		
Standard timebase	<0.1 ppm/month	
Optional oven timebase	<0.03 ppm/month	

counter) at some point in time, then one month later it may be off by as much as 0.1 ppm. For a frequency of 100 MHz, the maximum error would be 10 Hz. One year later, the frequency could be off as much as 12×0.1 ppm = 1.2 ppm. This results in an error of 120 Hz on a 100-MHz signal. Similar calculations can be made for the effect of temperature stability.

6.10 TIME INTERVAL ANALYZER

Most frequency measurements are made on stationary sine waves so the frequency is not varying over time. In the case of pulsed RF, the sine wave is present only part of the time so the counter must measure the frequency during the pulse ON time. This is a simple example of a signal which is not one stable frequency, but changes over time. Other examples are even more complex. For instance, a voltage controlled oscillator (VCO) circuit might be designed to sweep in frequency, starting at f_1 and sweeping linearly to f_2 (Figure 6-10). With such a measurement, a normal frequency counter would be confused and would give a reading that represents the average frequency over some segment of time.

To handle complex signals such as the sweeping VCO, the TIME INTERVAL ANALYZER (also known as the MODULATION DOMAIN ANALYZER or FREQUENCY AND TIME INTERVAL ANALYZER) was invented (Figure 6-11). This class of instrument measures the instantaneous frequency as a function of time. This may seem confusing at first, since electrical engineers are more used to waveforms being represented by voltage as a function of time. Engineers familiar with frequency domain measurements made with spectrum analyzers (Chapter 8) view signals as voltage versus frequency. But frequency versus time? This is something new. Returning to the case of the sweeping VCO, it is easy to understand the need for such an instrument. How could an engineer tell how linear the VCO sweep is? While it is theoretically possible to use an oscilloscope display to capture the voltage versus time information and convert it to frequency versus time, the

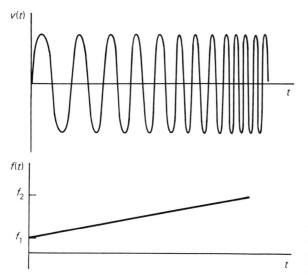

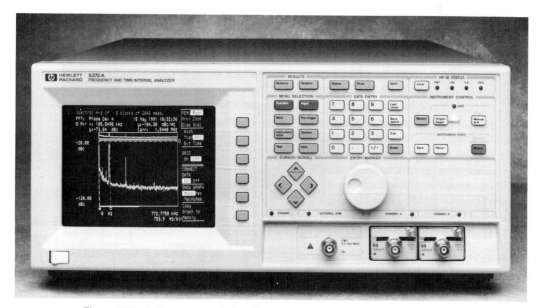

Figure 6-10 The frequency of an oscillator which sweeps from f_1 to f_2 can be tracked as a function of time by a modulation domain analyzer.

memory size of the scope must be very large and the timing resolution must be very fine. A more direct and efficient way to make this measurement is to use an instrument which measures the time between zero crossings on the waveform and plots these as a function of time. Thus, a time interval analyzer can plot the waveform's period versus time, frequency versus time, phase versus time, or time interval versus time.

Figure 6-11 A frequency and time interval analyzer. (Photo courtesy of Hewlett-Packard Company.)

The time interval analyzer is essentially a frequency counter that can produce a real-time frequency reading which is plotted as a function of time. The key component in the modulation domain analyzer is the ZERO DEAD TIME (ZDT) COUNTER, which can track the input signal's frequency and output the resulting value without missing a period of the signal. Otherwise, the counter would miss portions of the waveform, giving a discontinuous frequency plot.

The sweeping VCO represents a common use of the modulation domain analyzer. Another application is the measurement of time jitter and frequency modulation on an otherwise stable carrier. Some spread-spectrum radio transceivers use frequency hopping techniques to reduce the susceptibility to jamming and other interference. Time interval analyzers are very effective at tracking the frequency of the receiver's local oscillator as it jumps from place to place.

REFERENCES

"Fundamentals of Electronic Counters." Hewlett-Packard Company, Application Note 200, Publication Number 5952-7506, July 1978, Palo Alto, CA.

"Fundamentals of Microwave Frequency Counters." Hewlett-Packard Company, Application Note 200-1, Publication Number 5952-7484, September 1988, Palo Alto, CA.

HELFRICK, ALBERT D., and WILLIAM D. COOPER. *Modern Electronic Instrumentation and Measurement Technique.* Englewood Cliffs, NJ: Prentice-Hall, Inc., 1990.

OLIVER, BERNARD M., and JOHN M. CAGE. *Electronic Measurements and Instrumentation.* New York: McGraw-Hill Book Company, 1971.

"The Modulation Domain, Powerful New Analysis of Frequency, Phase and Time." Hewlett-Packard Company, Publication Number 5952-7997D, October 1989.

chapter seven

Circuits for Electronic Measurements

This chapter presents some important circuit concepts that relate to electronic measurements. Electronic circuits are usually the devices that are to be characterized or measured. The voltages or currents of some circuit that has already been designed and built often are measured to evaluate or repair the circuit. In other cases, it may be necessary to design and construct a simple circuit to perform the measurement. The circuits that aid electronic measurement generally fall into two categories: circuits which produce the measurement parameter (voltage or current) and circuits which condition a voltage or current that already exists.

7.1 RESISTANCE MEASUREMENT—INDIRECT METHOD

Modern ohmmeters (or multimeters) provide a convenient and accurate means of measuring resistance values. For most applications, this is the fastest and easiest way to make resistance measurements. There are other situations, however, where the ohmmeter is not capable of making the desired resistance measurement. Recall from Chapter 2 that the ohmmeter requires that all power sources be removed from the circuit under test. For measuring devices such as the input and output resistance of an amplifier, this is not possible. In fact, there is no single resistor to be measured since the input and output resistance of an amplifier is the equivalent resistance looking into the circuit. This equivalent resistance depends on the active components (usually transistors) and bias circuits inside the amplifier. The circuit must have power applied to measure the input or output resistance correctly.

One way to measure a resistor is to use an indirect method, which measures circuit parameters other than the unknown resistance and then computes from them the unknown value. Consider our old friend the voltage divider, shown in Figure 7-1. The voltage divider equation for this circuit is

$$V_L = V_S \frac{R_L}{R_S + R_L}$$

Rearranging, R_L and R_S can be found in terms of the other values:

$$R_L = \frac{V_L R_S}{V_S - V_L}$$

$$R_S = R_L \left(\frac{V_S}{V_L} - 1 \right)$$

So, for instance, if V_L, V_S, and R_S are known, the value of R_L can be determined.

7.1.1. Output Resistance

Suppose the output resistance of a signal source is to be measured. An ohmmeter cannot be used since the source must be powered up during the measurement. The source's output can be modeled as a Thévenin equivalent circuit as shown in Figure 7-2a. The value of V_S can be determined by measuring the output voltage of the source under open-circuit conditions. (Of course, the level of the source depends on its control settings, but for a given instrument setting, V_S will be constant.) A known load resistor can then be connected to the output of the source, and V_L can be measured (Figure 7-2b). V_L should always be less than or equal to V_S, due to the loading effect. Some experimentation may be required to determine a suitable value for the load resistor. If R_L is too large, indicated by V_L being very close to or the same as V_S, then the circuit is not being loaded enough. If R_L is too small, indicated by V_L being only a small fraction of V_S, the source may become too heavily loaded and may no longer

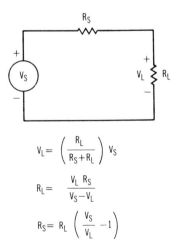

$$V_L = \left(\frac{R_L}{R_S + R_L} \right) V_S$$

$$R_L = \frac{V_L R_S}{V_S - V_L}$$

$$R_S = R_L \left(\frac{V_S}{V_L} - 1 \right)$$

Figure 7-1 The voltage divider circuit can be used for measuring the value of an unknown resistance.

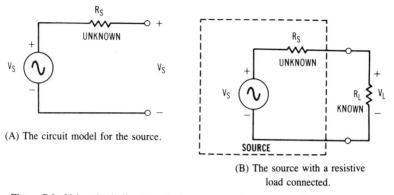

(A) The circuit model for the source.

(B) The source with a resistive
load connected.

Figure 7-2 Using the indirect method to measure the output resistance of a signal
source. (a) The circuit model for the source. (b) The source with a resistive load
connected.

operate within specification. In general, with a suitable value for R_L, V_L should be no
more than 90 percent of V_S and no less than 50 percent of V_S. Loading the source such
that V_L is 50 percent of V_S eliminates the need for any further calculation since, under
this condition, R_S equals R_L.

    This technique works on a variety of devices that can be modeled as a voltage
source with an internal series resistance. Examples are signal sources, batteries, and
(the output of) amplifiers. The same concepts apply whether the voltages involved
are DC or AC. In the AC case, it is important to note that the output impedance must
be resistive. Fortunately, this is true for many of the output impedances that are
measured. As the frequency of the measurement is increased, capacitive and inductive
effects become more significant, so this technique is most useful at audio frequencies.
Some devices (batteries, for example) will not tolerate heavy loading and must be
measured with as large a load resistor as possible.

Example 7-1.

    The output resistance of an amplifier is to be determined using the indirect method. A
sine wave source was connected to the input, causing a 2-volt RMS AC voltage at the
output under no load conditions (Figure 7-3). A 500-Ω load resistor connected to the
output caused the output voltage to drop to 1.5 volts RMS. What is the output resistance
of the amplifier?

    V_S is equal to the open circuit (no load) voltage, so $V_S = 2$ volts RMS. V_L was measured
at 1.5 volts RMS, and R_L is 500Ω. (Note that RMS values are used for V_S and V_L.
Zero-to-peak or peak-to-peak could also be used as long as they are used consistently.)

$$R_S = 500\left(\frac{2}{1.5} - 1\right) = 167\Omega$$

7.1.2 Input Resistance

Another problem is measuring the input resistance of devices, such as amplifiers and
filters. The input of an amplifier can be modeled as a single resistor, as shown in Figure
7-4a. If a source is connected to such an input, a voltage divider results, with the input

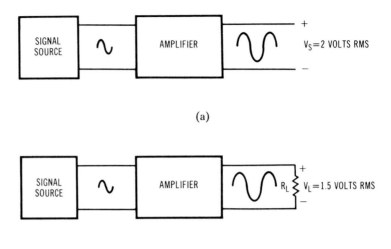

(a)

(b)

Figure 7-3 The output resistance of an amplifier is to be measured. (a) A source is connected to the input of the amplifier to produce an AC voltage at the amplifier's output. (b) The output amplifier is loaded with a resistor, and the output voltage is measured.

resistance of the amplifier acting as the load resistor (Figure 7-4b). With such a connection, neither R_S nor R_L can be varied. This is acceptable if a suitable amount of loading occurs (V_L being 50 percent to 90 percent of V_S). More often, an additional resistance is inserted in series with the output resistance of the source (Figure 7-4c). This resistor can be chosen (with some experimentation) to provide a reasonable amount of loading. For calculation purposes, the new R_S is the source's output resistance plus the added resistor. The open-circuit voltage of the source (V_S) and the loaded voltage (V_L) are measured, and R_L is computed. When using an additional series resistor, it is important to measure V_L across R_L (and not across the output of the source). Again, the circuits being measured must be resistive for this method to work properly. The input capacitance of the amplifier can be significant, especially when the input has a very high impedance. Generally, this method is practical for input impedances up to 100 kΩ and for frequencies less than 20 kHz.

Example 7-2.

The input resistance of an amplifier is to be measured using the indirect method. A sine wave source with open-circuit voltage 0.2 volts zero-to-peak and output resistance 600Ω is connected to the amplifier using a 10-Ω resistor (as shown in Figure 7-4c). The voltage at the input of the amplifier is 0.12 volts zero-to-peak. Determine the input resistance of the amplifier.

R_S, for our calculation, is the sum of the source's output resistance and the additional 10-kΩ resistor. $R_S = 10.6\ k\Omega$, $V_S = 0.2$ volts, and $V_L = 0.12$ volts.

$$R_L = \frac{0.12 \times 10.6\ k}{0.2 - 0.12} = 15.9\ k\Omega$$

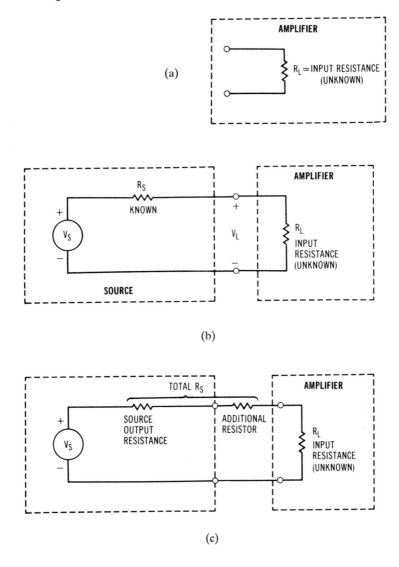

Figure 7-4. The indirect method can be used to determine the input resistance of a device.

7.2 BRIDGE MEASUREMENTS

Another method of measuring resistance (and other parameters such as inductance and capacitance) is the BRIDGE CIRCUIT. Resistance measuring methods previously discussed all require accurate voltage and/or current measurements. The bridge method requires only a null indication (minimum voltage reading) and precise reference resistors. Although the instrument user may construct a bridge circuit for measurement use, it is more likely that the user will encounter a commercially made measurement bridge. High-quality ohmmeters have largely replaced resistance bridges, but impedance bridges are still commonly used for measuring complex impedances, inductances, and capacitances. The general concepts concerning bridge measurements will be discussed here to familiarize the reader with the technique.

7.2.1 The Wheatstone Bridge

The Wheatstone bridge is the simplest and most easily understood measurement bridge (Figure 7-5). The measurement device (traditionally, a meter) is "bridged" across the circuit. The meter will read zero when the following condition is satisfied:

$$\frac{R_1}{R_2} = \frac{R_4}{R_3}$$

(This can be shown by treating the pairs R_1, R_2 and R_3, R_4 as voltage dividers driven by V_S. V_M is the difference between the output voltages of the voltage dividers.)

Suppose R_1 equals R_2 and R_3 is the unknown resistor to be measured. The meter will read zero (null indication) when

$$1 = \frac{R_4}{R_3} \quad \text{or} \quad R_3 = R_4$$

If R_4 is a variable precision resistor, its value can be adjusted until the meter reads zero. At this setting, the value of the unknown resistor is the same as the value of R_4. This assumes that R_4 has a calibrated dial connected to it so that its value can easily be determined. R_4 may be implemented using a bank of switchable resistors with or without additional variable resistors.

R_1 and R_2 were assumed to be equal in the preceding discussion. For additional flexibility, they may be made switchable. The null condition will occur whenever the ratio of R_1 and R_2 equals the ratio of R_3 and R_4. So changing the ratio of R_1 and R_2 alters the ratio of R_3 and R_4 that will produce the null condition. The measurement range of the bridge can be extended by using selectable values for R_1 or R_2.

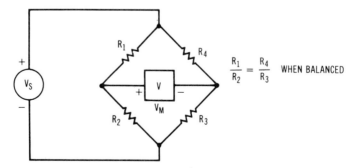

Figure 7-5 The Wheatstone bridge circuit is capable of comparing ratios of resistors very accurately. The meter displays its minimum value when the bridge is properly balanced.

7.2.2 Impedance Bridges

The concept of the Wheatstone bridge can be extended to facilitate impedance measurement. Such a bridge may be used to measure the values of capacitors and inductors. A variety of bridge configurations have been developed to accomplish this, but all are variations on the concept of the resistive bridge. One or more of the resistors in the bridge are either supplemented or replaced by inductors and capacitors. The unknown impedance is one arm of the bridge circuit, and one or more of the other arms are adjusted until the null condition is reached. Then the value of the unknown impedance can be inferred from the values of the components in the other arms in the bridge. Of course, since this is an impedance measurement, an AC (sine wave) source is used for V_S.[1]

7.3 RL AND RC CIRCUITS

In most cases, resistance measurements are simple as long as a good-quality multimeter or ohmmeter is available. Measuring inductors and capacitors, however, can be difficult. Although a specialized instrument such as an impedance bridge can be used to measure inductance and capacitance, there are ways of measuring inductors and capacitors with more common lab equipment.

7.3.1 Step Response

A resistor-capacitor (RC) circuit and a resistor-inductor (RL) circuit are shown in Figure 7-6. These two circuits have some very similar properties that can be used to determine the value of either the inductor or capacitor in the circuit. If the input (V_S) of the circuit is abruptly stepped from 0 volts to some positive value, the output voltage (V_O) rises in an exponential manner (Figure 7-7).

The mathematical expression for the step response of the circuits shown is

$$v_o(t) = V_S \left(1 - e^{-t/\tau}\right) \qquad \text{for } t \geq 0$$

The output voltage that results from a voltage step at the input is called the STEP RESPONSE. Notice that the step response does not immediately reach the final V_S level, but rises in a more sluggish manner. Theoretically, it will take an infinite amount of time for v_o to settle to its final value (which is the same as the final value of V_S). The standard method for describing how quickly the circuit responds is the TIME CONSTANT of the circuit.

In one time constant, the step response reaches 63.2 percent of its final value.

$$v_o(\tau) = V_S \left(1 - e^{-\tau/\tau}\right)$$
$$v_o(\tau) = V_S \left(1 - e^{-1}\right) = 0.632 \, V_S$$

---

[1]For a more complete discussion of other bridge circuits, see Helfrick and Cooper (1990) or Jones and Chin (1991).

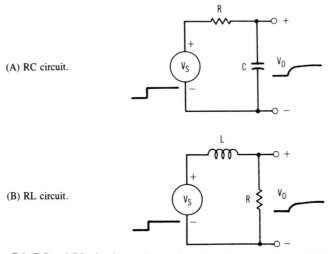

(A) RC circuit.

(B) RL circuit.

Figure 7-6 RC and RL circuits can be used to determine capacitor and inductor values by measuring the time constant of the circuit. (a) RC circuit. (b) RL circuit.

The time constant can be computed from the circuit values:

$$\tau = R \cdot C \quad \text{for the RC circuit}$$
$$\tau = L/R \quad \text{for the RL circuit}$$

An unknown capacitor or inductor whose value is to be determined can be connected to a known resistor in the appropriate circuit. (The resistor's value can be easily measured with an ohmmeter.) The time constant of the circuit is measured and the unknown component value computed.

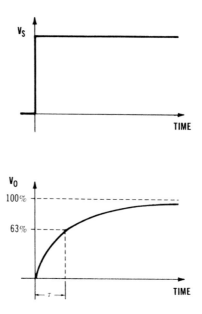

Figure 7-7 The step response input (V_S) and output (V_O) waveforms for the RC and RL circuits shown in Figure 7-6.

$$C = \frac{\tau}{R} \quad \text{for the RC circuit}$$

$$L = R \cdot \tau \quad \text{for the RL circuit}$$

Since the time constant is obviously a time domain parameter, an oscilloscope is required to measure it (Figure 7-8). For the step input, a switch connected to a voltage source might work, but a function generator is usually more practical. The function generator is set to output a fairly low-frequency square wave (about 100 Hz), which acts as a repetitive step voltage. The period of the square wave must be long enough to allow the circuit to settle (within the desired accuracy) to its final value before receiving the next voltage step (rising edge of the next square wave cycle). The amplitude of the square wave is fairly arbitrary, as long as the voltage is large enough to measure easily with the scope, and component ratings are not exceeded.

To a certain extent, the value of the resistor will depend on the value of the capacitor or inductor being measured. At first glance, this may seem like a real problem, since the goal is to measure the capacitor or inductor. However, a small amount of experimentation and experience will simplify the process. An experienced user can make a very rough estimate of the size of capacitor by its physical construction. For example, a polarized electrolytic capacitor is likely to be in 1 µF to 100 µF, while a small ceramic capacitor is more likely to be less than 1 µF. To use the oscilloscope to measure the time constant conveniently, it is recommended that the time constant be kept within the range of 10 msec to 10 µsec. A recommended resistance value to start with is 1 kΩ. The function generator's output resistance appears in series with the resistance and should be included in the calculations. Figure 7-9 shows the input and output voltages with a suitable time constant and oscilloscope setup.

Example 7-3

> The step response shown in Figure 7-9 resulted from an RC circuit with a 1-kΩ resistor and an unknown capacitor. The output impedance (resistance) of the source is 50 Ω. Determine the capacitor value.

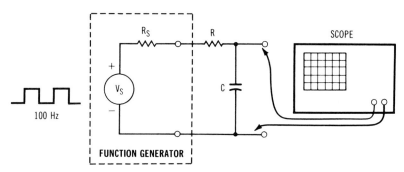

Figure 7-8 A function generator and an oscilloscope are used to measure the step response of this RC circuit. The function generator's output resistance is in series with R and will affect the time constant of the circuit.

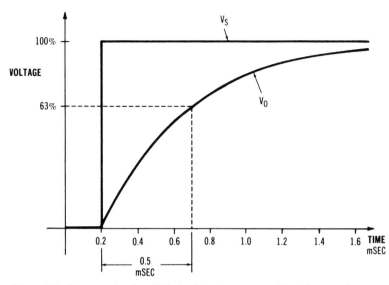

Figure 7-9 An example of an RC circuit's step response. Both input and output waveforms are shown. (See Example 7-3.)

From Figure 7-9, the step response takes 0.5 msec to reach 63 percent of its final value. Thus, the time constant is 0.5 msec. R is 1kΩ, but the output resistance of the function generator should also be included (in series).

$$C = \frac{\tau}{R} = \frac{0.5 \text{ msec}}{(1000 + 50)} = 0.476 \ \mu F$$

7.3.2 Frequency Response

These same circuits, except with a sine wave source driving them, can be used in the frequency domain to measure the value of a capacitor or inductor (Figure 7-10). The frequency responses of these two circuits are the same (Figure 7-11). Low-frequency signals are passed from the input to the output with little or no attenuation, while high-frequency signals are attenuated significantly. Hence, it forms a low-pass filter. Normally, the point at which the response has fallen 3 dB (relative to the response at low frequencies) is used to define the filter bandwidth. A loss of 3 dB corresponds to a reduction in output voltage to 70.7 percent of the original value.

$$f_{3dB} = \frac{1}{2\pi\tau}$$

where τ is the time constant of the circuit previously defined. (This is another example of how the time domain and frequency domain concepts are related.)

Although the 3-dB point is the classical way of defining where a low-pass filter rolls off, other points such as the 6-dB point may be used. A 6-dB reduction in voltage

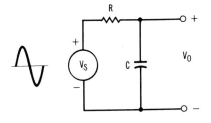

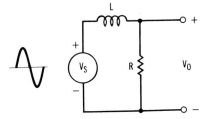

Figure 7-10 The frequency response of the RC and RL circuits can be used to measure a capacitor or an inductor.

corresponds to a 50 percent reduction in voltage. This will be a convenient number for measurement use.

$$f_{6dB} = \frac{\sqrt{3}}{2\pi\tau}$$

The equations relating the 3-dB and 6-dB frequencies to the circuit component values are summarized in Table 7-1.

To measure an inductor or capacitor, the unknown component is connected in an RC or RL circuit as appropriate. It is assumed that the resistor is known or can be accurately measured independently. The entire frequency response of the circuit does not need to be measured. It is sufficient to tune the sine wave source to a low frequency (usually 100 Hz or so), note the amplitude of the output voltage using an oscilloscope

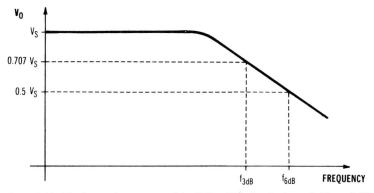

Figure 7-11 The frequency response of the RC and RL circuits shown in Figure 7-10.

TABLE 7-1 EQUATIONS RELATING THE CIRCUIT VALUES TO THE
FREQUENCY RESPONSE OF THE RC AND RL LOW-PASS CIRCUITS

RC Circuit	RL Circuit
$\tau = RC$	$\tau = \dfrac{L}{R}$
$f_{3dB} = \dfrac{1}{2\pi\tau} = \dfrac{1}{2\pi RC}$	$f_{3dB} = \dfrac{1}{2\pi\tau} = \dfrac{R}{2\pi L}$
$f_{6dB} = \dfrac{\sqrt{3}}{2\pi\tau} = \dfrac{\sqrt{3}}{2\pi RC}$	$f_{6dB} = \dfrac{\sqrt{3}}{2\pi\tau} = \dfrac{\sqrt{3}\,R}{2\pi L}$
$C = \dfrac{1}{2\pi R f_{3dB}} = \dfrac{\sqrt{3}}{2\pi R f_{6dB}}$	$L = \dfrac{R}{2\pi f_{3dB}} = \dfrac{\sqrt{3}\,R}{2\pi f_{6dB}}$

or meter, and then increase the frequency of the sine wave source until the output voltage drops 3 or 6 dB. The 6-dB point is probably more convenient, occurring when the voltage drops to half of its original value. This technique depends on the output of the source being constant with changes in frequency. (If the source amplitude is not constant with frequency, the frequency response must be calculated at each frequency point as the output voltage divided by the input voltage.) Although the entire frequency response does not have to be measured, it is good practice to trace out the response mentally when adjusting the source frequency. There should be no sharp peaks or dips in the response, just a gradual roll-off with increasing frequency. The 3-dB or 6-dB frequency may be determined from the frequency control of the source. Or for more accuracy, a frequency counter can be used to measure the source frequency. After the 3-dB or 6-dB frequency is measured, the value of the inductance or capacitance is calculated using the appropriate equation in Table 7-1.

Example 7-4

The following gain measurements were made on an RL low-pass circuit having a 500-Ω resistor. The output impedance of the source is 600 Ω. Determine the value of the inductor.

Freq (Hz)	Gain
2,000	0.96
4,000	0.87
6,000	0.76
8,000	0.65
10,000	0.57
12,000	0.50
14,000	0.44

The gain is 0.5 at 12 kHz, so the 6-dB equation is the most convenient in this case. Using the appropriate equation from Table 7-1;

$$L = \frac{\sqrt{3}\,R}{2\pi f_{6dB}}$$

R in the equation must include both the resistor value and the output impedance of the source. $R = 500 + 600 = 1100 \, \Omega$.

$$L = \frac{1.732 \times 1100}{2 \times 3.142 \times 12{,}000} = 25.3 \text{ mH}$$

7.4 RESONANT CIRCUITS

RESONANCE is another property of inductors and capacitors that can be used to measure their value. At resonance, the impedance of an inductor and capacitor exactly cancel, creating a sharp maximum or minimum in a circuit's response. Consider Figure 7-12a where a parallel LC network is shown. The frequency response of this circuit is a bell-shaped response that peaks at f_0, the resonant frequency. At this frequency, the parallel LC circuit reaches its highest impedance (ideally, infinite). The loading effect (of the LC circuit upon the source and resistor) is minimized at this frequency. Thus, the output is at a maximum. The resistor value should be set experimentally, starting with about 10 kΩ. A larger resistor causes a sharper resonance (more peaked response) at the expense of a lower output voltage. Generally, the sharper the resonance, the better the measurement.

The resonant frequency depends only on the L and C values:[2]

$$f_0 = \frac{1}{2\pi\sqrt{L \cdot C}}$$

Figures 7-12b and 7-12c show a series version of the LC circuit. Note that no resistor is required in this case. The output voltage is maximum at resonance for both these circuits, but the shape of the curve is slightly different for each one. The equation for the resonant frequency is valid for both series and parallel LC networks. The sharpness of the resonance is dependent mainly on the quality of the inductor and the output impedance of the source (the lower the output impedance, the sharper the resonance).

For measuring either an inductor or a capacitor, the value of the other component in the circuit must be known. The frequency of the source is adjusted until the output voltage peaks—this is the resonant frequency. The value of the unknown component can be calculated using the following equations:

$$L = \frac{1}{4\pi^2 f_0^2 C}$$

$$C = \frac{1}{4\pi^2 f_0^2 L}$$

The choice of which circuit to use depends on the particular circuit components and is best determined experimentally. The circuit that gives the sharpest resonance is the preferred one. The known component should be chosen to keep the resonant

---

[2]See Appendix D for a more complete analysis of the resonant frequency concept.

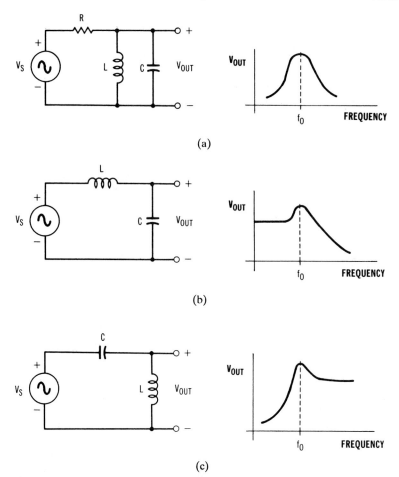

Figure 7-12   LC resonant circuits for measuring the value of inductors and capacitors. (a) Parallel LC circuit. (b) Series LC with voltage taken across the capacitor. (c) Series LC with the voltage taken across the inductor.

frequency below 10 MHz, where possible. Otherwise, stray capacitance and other parasitic effects will disturb the circuit's response. It is important to be wary of circuit behavior that does not agree with the frequency response curves in Figure 7-12. Multiple resonances are a very common occurrence, but they can usually be ignored if they are at a much higher frequency than the desired resonance. Responses that do not have a distinct peak are suspect—try another one of the circuit configurations.

Example 7-5.

The resonant frequency of the parallel LC circuit (Figure 7-12a) is found to be 18 kHz. The output impedance of the source is 50 Ω and the resistor value is 10 kΩ. If the inductor is 25 μH, what is the value of the capacitor?

Using the equation for the value of the capacitor;

$$C = \frac{1}{4\pi^2 f_0^2 L}$$

$$= \frac{1}{4(3.142)^2(18000)^2 25\mu H} = 3.13 \ \mu F$$

7.5 DIODE MEASUREMENT CIRCUIT

Although a diode can usually be tested using either an ohmmeter or an oscilloscope (and curve tracer circuit), it is also possible to measure a diode using a voltmeter and power supply. While the ohmmeter approach is usually the most convenient, it may not be possible to determine the forward voltage drop of the diode using this technique. The curve tracer circuit gives a more complete characterization of the diode, but it requires the use of an oscilloscope.

The circuit shown in Figure 7-13a can be used to measure the voltage drop across the diode when forward biased (forward voltage drop). The circuit supplies a positive DC voltage that forces current through the diode in the forward direction (assuming the diode is functioning properly). The voltage across the diode can be measured using a voltmeter or scope. V_S should be set to be significantly larger than the expected forward drop of the diode, but not so large that the diode could be damaged by excessive voltage or current. R_S limits the maximum amount of current that can be forced through the diode. $V_S = 5$ volts and $R_S = 5 \ k\Omega$ will forward bias most common diodes without causing any damage. The forward drop of a functioning diode depends on the type of diode, but it is typically less than 1 volt. (Table 7-2 shows the values of forward voltage drop of various types of diodes.) A standard silicon diode has a forward drop of about 0.6 volt, and a germanium diode has a forward drop of about 0.3 volt. If the measured diode voltage is very close to V_S, then the diode is open circuited and is not functioning properly.

In the reverse direction, the diode should act like an open circuit. This can be verified by reversing the direction of the diode as shown in Figure 7-13b. The diode voltage should be approximately the same as V_S.

A Zener diode will behave like a regular diode until a large reverse-biased voltage is applied to it. Under such conditions, the Zener diode will limit its voltage (in the reverse direction) at its ZENER VOLTAGE. Figure 7-13c shows how the test circuit can be used to measure the Zener voltage of a Zener diode. V_S must be larger than the Zener voltage being measured. Note the polarity of the Zener diode—it is being measured in the reverse direction.

7.6 TERMINATIONS

Many circuits that are encountered in electronic measurement require that they be connected to a known impedance or resistance for proper circuit operation. This may be the result of having a constant input and output impedance (Z_0) throughout the

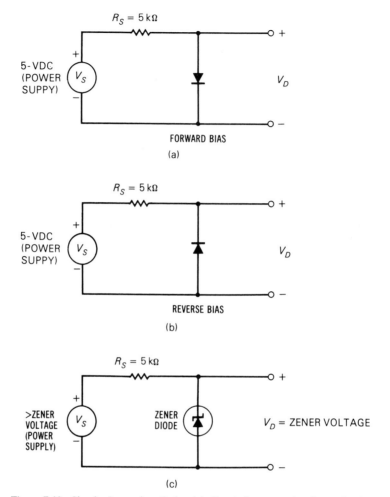

Figure 7-13 Circuits for testing diodes. (a) Circuit for measuring forward voltage drop. (b) Circuit for testing the reverse-biased condition. (c) Circuit for measuring the Zener voltage of a Zener diode.

TABLE 7-2 TYPICAL VOLTAGE DROPS FOR FORWARD-BIASED DIODES.

Diode Type	Forward Voltage
Germanium	0.25–0.4 volts
Standard silicon	0.60–0.7 volts
Schottky barrier	0.4–0.5 volts
Hot carrier	0.35–0.5 volts
Light emitting (LED)	1.5–2.5 volts

system where each piece of the system expects to be loaded by that particular impedance (50 ohms, 75 ohms, etc.). For the measurement to be valid, these same loading requirements must be maintained. One way to accomplish this is to use instruments which are inherently the proper impedance. Another way is to use a TERMINATION.

A termination (or load) is one of the simplest circuits used in electronic instrumentation, but an important one. A termination is nothing more than a resistor, usually packaged with convenient connectors. This resistor can be connected to a circuit under test to provide the proper load impedance. If the termination has two separate connection ports, then it is a FEEDTHROUGH TERMINATION (Figure 7-14). This type of termination is intended to be used with one side connected to the circuit under test and the other side connected to the measuring instrument. If the measuring instrument has a high input impedance (compared to the termination), then the circuit under test is essentially loaded by the termination resistance. (Remember, a small resistance in parallel with a very large resistance is roughly equivalent to just the small resistance.)

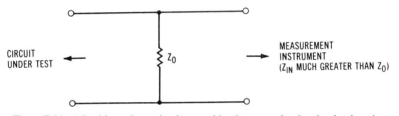

Figure 7-14   A feedthrough termination provides the proper load to the circuit under test, while simultaneously allowing a high-impedance instrument to be connected.

7.7 ATTENUATORS

An ATTENUATOR is sometimes necessary to reduce the level of the signal being measured. This may be due to the signal being so large that, unless it is reduced, a sensitive instrument will be overloaded. Attenuators also provide a certain amount of isolation between instruments, reducing any measurement interaction. Commercially built attenuators are available, including well-calibrated adjustable attenuators, but it is often more convenient to build one.

Three types of attenuators will be discussed:

1. *Voltage divider attenuators*: attenuators intended to be driven by low-impedance sources and loaded by high-impedance circuits or instruments. This definition removes the loading effect due to devices external to the attenuator.

2. Z_0 *Attenuators*: Attenuators intended to be driven by sources with a known system impedance (Z_0) and loaded by circuits or instruments with the same system impedance (Z_0).

3. *Impedance matching attenuators*: Attenuators intended to be driven by a particular source impedance (Z_1) and loaded by a different impedance (Z_2).

7.7.1 Voltage Divider

In the first case, the attenuator is simply a resistive voltage divider. Again, the loading effects external to the voltage divider itself have been removed so the circuit operation is straightforward. Figure 7-15 shows the voltage divider attenuator and its associated design equations. First, the sum of $R_1 + R_2$ is chosen arbitrarily. This sum ($R_1 + R_2$) is the resistive load that will be on the source so it should not be so small as to overload the source. (Typically, 1 kΩ to 10 kΩ is reasonable.) The values for R_1 and R_2 can be calculated from the equations in Figure 7-15. Some typical values for the case where $R_1 + R_2 = 1$ kΩ are tabulated in Table 7-3. If a larger or smaller resistive load is required, the values in Table 7-3 may be multiplied by a scale factor. As long as the same scale factor is used for both R_1 and R_2, the attenuation remains the same.

Example 7-6.

Determine the resistor values for a voltage divider attenuator whose output voltage is one-tenth of its input voltage. The attenuator should provide no less than a 10-kΩ load to the circuit driving it.

To meet the loading requirement, let $R_1 + R_2 = 10$ kΩ. The voltage gain required in the voltage divider is

$$G = \frac{V_{\text{OUT}}}{V_{\text{IN}}} = 0.1$$

$$R_2 = G(R_1 + R_2) = 0.1(10 \text{ k}\Omega) = 1 \text{ k}\Omega$$
$$R_1 = (1 - G)(R_1 + R_2) = 0.9(10 \text{ k}\Omega) = 9 \text{ k}\Omega$$

Alternatively, Table 7-3 shows a 20-dB attenuator (linear voltage gain equals 0.1) with $R_1 = 900$ and $R_2 = 100$. But this is for the case $R_1 + R_2 = 1$ kΩ which will not meet the loading requirement. These values can be scaled by a factor of 10 to provide an attenuator with $R_1 + R_2 = 10$ kΩ.

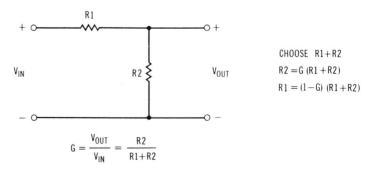

$$G = \frac{V_{\text{OUT}}}{V_{\text{IN}}} = \frac{R2}{R1 + R2}$$

Figure 7-15 The voltage divider can be used as an attenuator when its output is to be loaded by a high impedance.

$R_1 = 10 \times 900 = 9\text{ k}\Omega$

$R_2 = 10 \times 100 = 1\text{ k}\Omega$

which is the same answer previously calculated.

TABLE 7-3 TABLE OF VOLTAGE DIVIDER ATTENUATOR VALUES
FOR THE CASE $R_1 + R_2 = 1$ KΩ. R_1 AND R_2 MAY BE SCALED
AS NECESSARY WITHOUT CHANGING THE GAIN OF THE
ATTENUATOR.

dB	Voltage Gain linear	R1	R2
−1	0.891	109	891
−2	0.794	206	794
−3	0.708	292	708
−4	0.631	369	631
−5	0.562	438	562
−6	0.501	499	501
−10	0.316	684	316
−20	0.100	900	100
−30	0.032	968	32
−40	0.010	990	10

7.7.2 Z_0 Attenuators

For systems that have a particular input and output impedance, the attenuator circuits shown in Figure 7-16 can be used. These circuits are designed to be loaded by the same impedance (Z_0) on each end. They are also symmetrical devices—even though an input and output are labeled, they can be reversed in use. For consistency with the voltage divider attenuator, the design equations in Figure 7-16 use voltage gain. Often attenuators are specified in terms of power gain or loss. Since these particular attenuators have the same impedance at both ends, the voltage gain and power gain are the same WHEN EXPRESSED IN DECIBELS. Figures 7-16a and 7-16b show unbalanced and balanced T attenuators, respectively. Notice that the unbalanced version connects the ground side of the input directly to the ground side of the output. This is not usually a concern if both sides are grounded anyway, but it may be a problem where floating inputs or floating outputs are used (such as on a source, power supply, oscilloscope, etc.). The balanced version of the attenuator solves this problem at the expense of additional resistors. (Note that the R_1 resistors are half the unbalanced value.) Similarly, unbalanced and balanced versions of the Pi attenuator are shown in Figures 7-16c and 7-16d. In operation, the T and Pi attenuator circuits are equivalent. The choice between the two is usually made based on the practicality or convenience of the resistor values.

Resistor values have been tabulated in Table 7-4 for attenuators with Z_0 equal to 50 Ω. The values may be scaled to produce attenuators with other Z_0

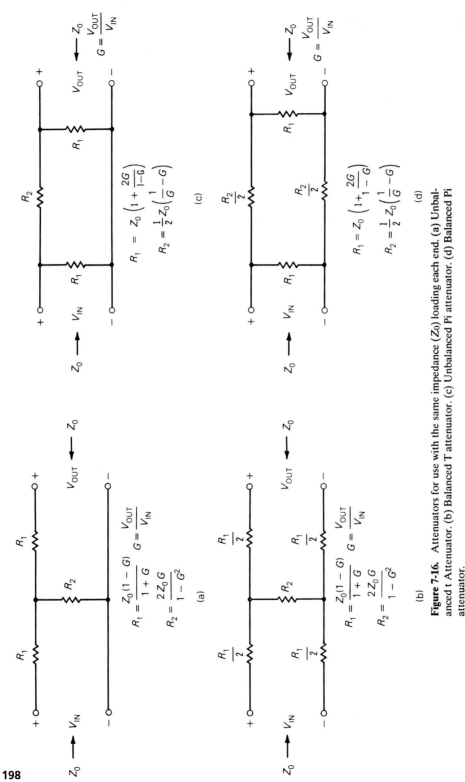

Figure 7-16. Attenuators for use with the same impedance (Z_0) loading each end. (a) Unbalanced t Attenuator. (b) Balanced T attenuator. (c) Unbalanced Pi attenuator. (d) Balanced Pi attenuator.

TABLE 7-4. RESISTOR VALUES FOR T AND PI ATTENUATORS WITH Z_0 EQUAL TO 50 OHMS. THE VALUES CAN BE SCALED TO PRODUCE ATTENUATORS WITH OTHER Z_0 VALUES.

	T Attenuator		
Voltage Gain			
dB	Linear	R_1	R_2
−1	0.891	2.875	433.3
−2	0.794	5.731	215.2
−3	0.708	8.549	141.9
−4	0.631	11.31	104.8
−5	0.562	14.00	82.24
−6	0.501	16.61	66.93
−10	0.316	25.97	35.13
−20	0.100	40.90	10.10
−30	0.032	46.93	3.165
−40	0.010	49.00	1.000

	Pi Attenuator		
Voltage Gain			
dB	Linear	R_1	R_2
−1	0.891	869.5	5.769
−2	0.794	436.2	11.61
−3	0.708	292.4	17.61
−4	0.631	220.9	23.84
−5	0.562	178.4	30.39
−6	0.501	150.4	37.35
−10	0.316	96.24	71.15
−20	0.100	61.11	247.5
−30	0.032	53.26	789.7
−40	0.010	51.01	2499.

values. For instance, a 75-Ω attenuator would use resistance values multiplied by 75/50, or 1.5.

Example 7-7.

Determine the resistor values for a 10-dB attenuator having both ends loaded by 600 Ω. We will arbitrarily choose the Pi attenuator shown in Figure 7-16c. The desired loss is 10 dB, which is a gain of −10 dB. Solving for G (not in decibels),

$$G_{dB} = 20 \log G$$

$$G = 10^{G_{dB}/20} = 10^{-10/20} = 0.3162 \qquad \text{(voltage ratio)}$$

$$R_1 = Z_0 \left[1 + \frac{2G}{1 - G} \right] = 600 \left[1 + \frac{2 \times 0.3162}{1 - 0.3162} \right]$$

$$R_1 = 1155 \ \Omega$$

$$R_2 = \frac{1}{2} Z_0 \left(\frac{1}{G} - G \right) = \left(\frac{1}{2} \right) 600 \left(\frac{1}{0.3162} - 0.3162 \right)$$

$R_2 = 854 \ \Omega$

Or using Table 7-4, $R_1 = 96.24$ and $R_2 = 71.15$ for $Z_0 = 50 \ \Omega$. Since the desired Z_0 is 600 Ω, these values must be multiplied by $(600/50) = 12$.

$R_1 = 12 \times 96.24 = 1155 \ \Omega$

$R_2 = 12 \times 71.15 = 854 \ \Omega$

These values can be used for either an unbalanced or balanced Pi attenuator. (If the balanced version is chosen, R_2 is halved.)

7.7.3 Impedance Matching Attenuators

Although many attenuators are designed and used with the same impedance loading both ends of the device, attenuators can also be designed to work with different impedances at each end. In particular, attenuators can be used to match one impedance to another. Some loss will always be present when using an impedance matching attenuator, but the loss can be minimized. Attenuators that have the theoretical minimum amount of loss while matching two different impedances are called MINIMUM LOSS PADS. Figure 7-17 shows the minimum loss pad circuit diagram. Notice that Z_1 must be the larger of the two impedances. The equations for determining the resistor values and the loss are included in Figure 7-17. Care must be taken in interpreting the loss in dB, since the impedances at each end are not equal.

Minimum loss pads are useful when the measuring instrument input impedance does not match the output impedance of the circuit under test. For example, it may be desirable to measure a 75-Ω system with the 50-Ω instrument input. The 75-Ω system should be loaded by 75 Ω during the measurement. Connecting the 50-Ω input directly will alter the conditions of the measurement and may even cause improper circuit operation. The problem can be solved by inserting a 50-Ω to 75-Ω minimum loss pad between the circuit under test and the instrument input, thereby providing an appropriate load or match for both the 75-Ω system and the instrument. Of course, there will be a small amount of loss in the pad, so the measured values should be adjusted accordingly.

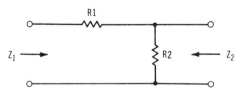

$$R1 = Z_1 \ \sqrt{1 - (Z_2/Z_1)}$$

$$R2 = \frac{Z_2}{\sqrt{1 - Z_2/Z_1}}$$

$$\text{LOSS (dB)} = 10 \ \text{LOG} \left[\frac{2 \, Z_1}{Z_2} - 1 + 2 \ \sqrt{\frac{Z_1}{Z_2} \left(\frac{Z_1}{Z_2} - 1 \right)} \right]$$

Figure 7-17 A minimum loss pad provides impedance matching between two different impedances at the expense of some signal loss. Note that Z_1 must be greater than Z_2.

Example 7-8.

Calculate the resistor values for a 50-Ω to 75-Ω minimum loss pad.

Using the equations from Figure 7-17, with $Z_1 = 75\ \Omega$ and $Z_2 = 50\ \Omega$

$$R_1 = Z_1 \sqrt{1 - \left(\frac{Z_2}{Z_1}\right)} = 75 \sqrt{1 - \left(\frac{50}{75}\right)} = 43.3\ \Omega$$

$$R_2 = \frac{Z_2}{\sqrt{1 - \left(\frac{Z_2}{Z_1}\right)}} = \frac{50}{\sqrt{1 - \left(\frac{50}{70}\right)}} = 86.6\ \Omega$$

The resulting circuit is shown in Figure 7-18.

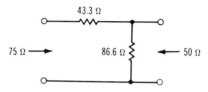

Figure 7-18 A minimum loss pad for connecting a 75-Ω system to a 50-Ω system. (See Example 7-8.)

7.8 POWER SPLITTERS AND COMBINERS

Another termination problem can arise with Z_0 systems when trying to drive two systems from a signal source or when combining two signals into one measuring instrument or circuit. Novice instrument users may attempt simply to connect all the devices in parallel to combine their outputs. In some cases this may prove satisfactory, but often the impedance matching and loading problems will be significant. A POWER SPLITTER (Figure 7-19) can be used to split one signal into two or to combine two signals into one. The delta power splitter consists of three resistors, each having the value Z_0, connected in a delta arrangement. The Y power splitter also has

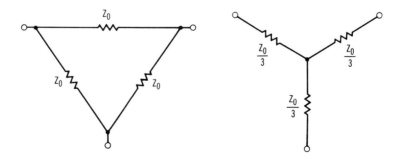

(A) The delta power splitter circuit.   (B) The Y power splitter circuit.

Figure 7-19 These two power splitter circuits are equivalent in operation. (a) The delta power splitter circuit. (b) The Y power splitter circuit.

three resistors, but they have the value $Z_0/3$ and are connected in a Y configuration. The two power splitter circuits are totally interchangeable. For proper operation, each connection or "port" of the power splitter must be loaded by a Z_0 impedance. Figure 7-20a shows the delta power splitter in use, with a signal source driving one port and the two other ports connected to Z_0 loads. As long as the power splitter is loaded with Z_0 at each port, the signal level delivered at the two loads is the same. The power splitter can just as easily be used to combine the output of two signal sources into one (Figure 7-20b). Note that each port is once again loaded by Z_0. The power splitter incurs a 6-dB loss (compared with connecting one source directly to one load) which must be accounted for when setting signal source levels or interpreting measured data.

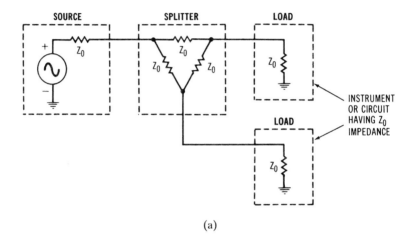

(a)

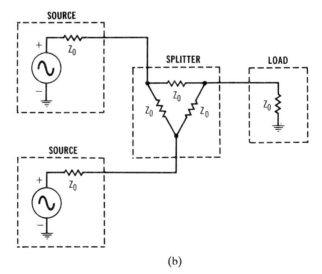

(b)

Figure 7-20 The power splitter can be used to split or combine signals. (a) Power splitter used for connecting one source to two Z_0 loads. (b) Power splitter used for connecting two sources to one Z_0 load.

7.9 MEASUREMENT FILTERS

Filters are circuits which pass certain frequencies while rejecting others. In a measurement situation, it may be desirable to insert a filter between the circuit under test and the measuring instrument to remove some unwanted signal or noise which threatens to degrade the measurement. For example, perhaps a low-frequency sine wave is accompanied by high-frequency noise. A low-pass filter could be used to remove the unwanted noise, while allowing the original sine wave to be measured. (As a side effect, this is equivalent to reducing the bandwidth of the measuring instrument.) For general-purpose measurement use, the 3-dB frequency should be chosen far enough away from the desired signal frequency so that it is *not* attenuated significantly. At the same time, the 3-dB frequency must not be too close to undesired frequency; otherwise, that signal will not be attenuated enough. Also, keep in mind that for some measurements, such as square wave testing, filtering at any frequency may not be appropriate.

Filter design is a complex subject and entire textbooks have been written about it. That level of discussion will not be duplicated here, but a few simple filter circuits can be outlined so that the casual user can make use of them. For more critical applications, it may be necessary to use more complex filter circuits.

Two types of filters will be discussed:

1. *High-impedance filters*: filters intended to be driven by low-impedance sources and loaded by high-impedance circuits or instruments. This definition essentially removes all loading effect due to devices external to the filter.

2. Z_0 *filters*: filters intended to be driven by sources with a known system impedance (Z_0) and loaded by instruments with the same system impedance (Z_0).

7.9.1 High-impedance Filters

The two RC circuits shown in Figure 7-21 can be used as filters when followed by high-impedance instruments. (The circuit in Figure 7-21a is the same RC circuit used for capacitance measurement earlier in the chapter.) The two circuits share the same

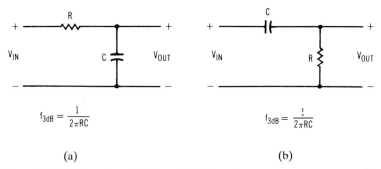

$$f_{3dB} = \frac{1}{2\pi RC}$$

(a)

$$f_{3dB} = \frac{1}{2\pi RC}$$

(b)

Figure 7-21   RC circuits used as filters in high-impedance measurement systems. (a) The RC low-pass filter. (b) The RC high-pass filter.

3-dB frequency, but Figure 7-21a has a low-pass frequency response, and Figure 7-21b has a high-pass frequency response. The frequency responses of the two circuits are plotted in Figure 7-22. The frequency axis is labeled in terms of the 3-dB frequency, since it is the basic design decision. Notice that the frequency response does not change abruptly at the 3-dB point but gradually rolls off.

Example 7-9.

Design a filter that will attenuate 60-Hz signals coupling from the power line onto the desired 2-MHz sine wave signal. The measuring instrument has a high-impedance input.

The design goal is to pass a higher frequency (2 MHz) and attenuate a low frequency (60 Hz). Therefore, a high-pass filter is required. Surveying our extensive collection of high-pass filters, Figure 7-21b is selected. From Figure 7-22, the high-pass characteristic has very little loss at 10 times the 3-dB frequency, so make the 3-dB frequency a factor of 10 less than the desired frequency (2 MHz). Again, looking at Figure 7-22, this should provide a considerable amount of rejection at 60 Hz, which is more than 3000 times less than the 3-dB frequency.

$$f_{3dB} = \frac{2 \text{ MHz}}{10} = 200 \text{ kHz}$$

$$= \frac{1}{2\pi RC}$$

$$R \cdot C = \frac{1}{2\pi f_{3dB}} = \frac{1}{2\pi(200 \text{ kHz})} = 7.96 \times 10^{-07}$$

Either R or C can be arbitrarily chosen, so choose $R = 1000 \ \Omega$.

$$C = \frac{7.96 \times 10^{-07}}{R} = 796 \text{ pF}$$

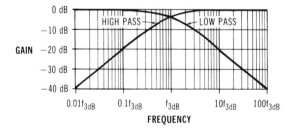

Figure 7-22 The frequency response of the low-pass and high-pass circuits from Figure 7-21. The frequency axis is plotted relative to the 3-dB frequency.

7.9.2 Z_0 Filters

The operation of the previous filter circuits will be disturbed if they are loaded by typical Z_0 values (50 ohms, 75 ohms, etc.). Figure 7-23 shows a low-pass and a high-pass circuit (along with their design equations) appropriate for use in terminated systems. Both circuits are classified as second-order Butterworth networks, a type of filter which has a flat-topped frequency response. The frequency response of the two

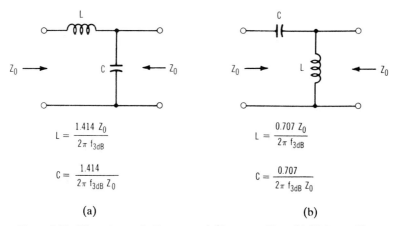

Figure 7-23 Filters for use in Z_0 systems. (a) Low-pass filter. (b) High-pass filter.

circuits are shown in Figure 7-24 with the frequency axis shown relative to the 3-dB frequency. The design of such a filter is relatively straightforward—the L and C values are determined by the design equation, depending on Z_0 and f_{3dB}. Both ends of the filter must be terminated in Z_0 for proper circuit operation.

Example 7-10

Design a low-pass filter with a 5-MHz 3-dB frequency for use in a 50-Ω system. Using the circuit and equations from Figure 7-23a,

$$L = \frac{1.414 Z_0}{2\pi f_{3dB}} = \frac{1.414 \times 50}{2\pi \times 5 \text{ MHz}} = 2.25 \ \mu\text{H}$$

$$C = \frac{1.414}{2\pi f_{3dB} Z_0} = \frac{1.414}{2\pi \times 5 \text{ MHz} \times 50} = 900 \text{ pF}$$

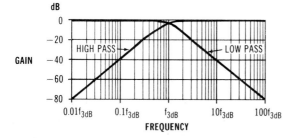

Figure 7-24 The frequency response of the low-pass and high-pass circuits from Figure 7-23. The frequency axis is plotted relative to the 3-dB frequency.

7.10 POWER SUPPLIES

Power supplies do not measure circuit parameters such as voltage and current, but instead supply circuits with DC voltage and current. Most electronic circuits require a DC voltage to operate. This can be supplied by a battery, a built-in power supply that changes AC voltage to DC voltage, or a DC bench supply. For portable applications, the battery is a good solution. For circuits that are used where AC power is available, having a built-in power supply makes sense. But when designing or testing a circuit, a bench DC supply will usually be the most convenient.

Power supplies may either be fixed or variable. As the name implies, a fixed power supply has a constant output voltage. An adjustment may be included on a fixed power supply to set the output voltage precisely. Since the adjustment range is usually only 5 or 10 percent around the nominal value, the supply is still considered to have a fixed voltage.

Variable supplies can be adjusted from the front panel to produce a wider range of voltages, typically 0 to 20 volts or more. This type of supply is more versatile since the voltage can be set to match the particular circuit requirement. In addition, the voltage can be varied around the nominal value during design and testing.

The circuit diagram for a simple power supply is shown in Figure 7-25. The AC line voltage (typically 120 volts RMS) is connected to the transformer which steps down the voltage to something closer to the final DC value. The diode (or rectifier) changes the AC sine wave voltage into a half sine wave. The filter capacitor smooths out the half sine to approximate a DC voltage. This DC voltage is not well controlled, so it is passed through a circuit called a VOLTAGE REGULATOR, which precisely controls the output voltage. This is one of the simplest types of power supplies, shown here to introduce briefly the steps necessary to change AC into DC. Practical power supplies are often much more complicated.

7.10.1 Power Supply Specifications

There will be imperfections in the DC voltage. The DC voltage may vary as the amount of current drawn from it changes. There may be a small amount of AC RIPPLE remaining riding on top of the DC (Figure 7-26). The AC ripple is a remnant of the AC line voltage which is not removed by the filter capacitor and voltage regulator. The ripple will be at the line frequency (usually 60 Hz in the United States) plus the harmonics of the line frequency (120 Hz, 180 Hz, etc.). Some power supplies use regulation techniques that will cause other frequencies to be present. In a quality power supply, all these ripple components will be low enough that they can usually be neglected (less than a few millivolts).

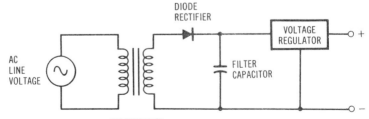

Figure 7-25 Schematic diagram for a simple DC power supply.

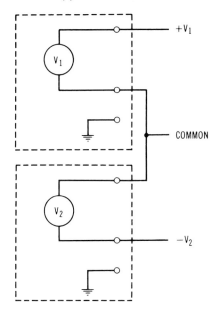

+V$_1$

COMMON

−V$_2$

Figure 7-30 Two single supplies can be connected to create a bipolar supply.

connected internally. The supplies shown are floating but can be grounded as needed by connecting the common terminal to the ground terminal. Figure 7-30 shows how two single power supplies can be connected to produce a bipolar power supply. At least one of the power supplies must be floating to avoid any conflicts in grounding.

A typical bench power supply is shown in Figure 7-31 and its specifications are listed in Table 7-5.

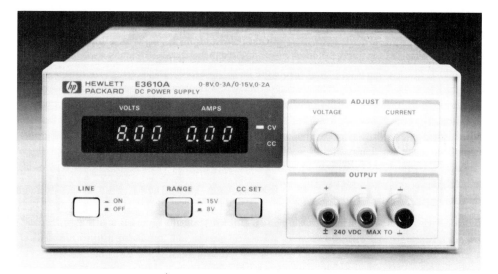

Figure 7-31 A typical bench power supply with digital readout. (Photo courtesy of Hewlett-Packard Company.)

TABLE 7-5 SPECIFICATIONS OF A TYPICAL SINGLE-OUTPUT
DC BENCH POWER SUPPLY.

Output Voltage: 0 to 15 volts, variable
Maximum Current: 3 amps
Load Regulation: 0.01% + 2 mV
Line Regulation: 0.01% + 2 mV
Ripple and Noise (10 Hz 2 10 MHz): 200 mV RMS 2mV p-p
AC Input: 115VAC ± 10% 47-63 Hz

REFERENCES

"DC Power Supply Handbook." Hewlett-Packard Company, Application Note 90B, Literature
Number 5952-4020, 1978, Palo Alto, CA.

HELFRICK, ALBERT D., and WILLIAM D. COOPER. *Modern Electronic Instrumentation and
Measurement Techniques*, Englewood Cliffs, NJ: Prentice-Hall, Inc., 1990.

JONES, LARRY D., and A. FOSTER CHIN. *Electronic Instruments and Measurements*, 2nd Ed.,
Englewood Cliffs, NJ: Prentice-Hall, Inc., 1991.

SCHWARZ, STEVEN E., and WILLIAM G. OLDHAM. *Electrical Engineering: An Introduction*, New
York: CBS College Publishing, 1984.

WILLIAMS, ARTHUR B. *Electronic Filter Design Handbook*, New York: McGraw-Hill Book
Company, 1981.

chapter eight

Frequency Domain
─────── *Instruments* ───────

In the preceding chapters, waveforms have been usually characterized and measured in the time domain. As demonstrated in Chapter 1, these same waveforms can be analyzed and measured in the frequency domain. This frequency domain representation is also referred to as the FREQUENCY SPECTRUM of the signal. Of primary interest is the SPECTRUM ANALYZER, which characterizes a waveform in the frequency domain the same way that an oscilloscope characterizes a waveform in the time domain. Closely related to the spectrum analyzer is the NETWORK AN-ALYZER, which is used to characterize the frequency response of electronic networks. Other instruments such as the DISTORTION ANALYZER and GAIN/PHASE METER also work in the frequency domain.

8.1 SPECTRUM ANALYZERS

Recall from Chapter 1 that a signal can be viewed in the time domain (Figure 8-1a) or the frequency domain (Figure 8-1b). The time domain plot is simply the instantaneous voltage plotted versus time. (Note that voltage terminology is being used, but the waveform could just as well be a current.) In the frequency domain, the vertical axis is still voltage, but the horizontal axis is now frequency. The two representations are consistent but different ways of looking at the same signal. In the frequency domain, the signal has various frequency components (spectral lines) that indicate the amount of energy at each frequency.

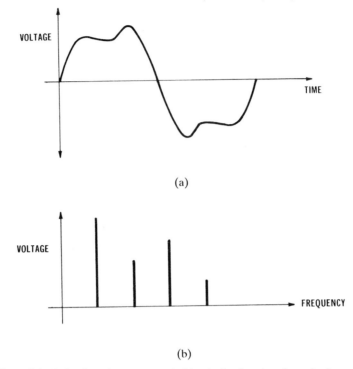

(a)

(b)

Figure 8-1 A signal can be represented either in the time domain or the frequency domain. (a) The time domain representation is voltage as a function of time. (b) The frequency domain representation is amplitude (voltage) as a function of frequency.

8.2 BANK-OF-FILTERS SPECTRUM ANALYZERS

One way to implement a spectrum analyzer is to use a large number of bandpass filters, each one tuned to a different frequency. Each filter removes all frequency components of the signal except the particular frequency that the filter was designed to measure. Figure 8-2 shows how each filter picks out a small section of the frequency axis to be measured. At the output of each filter there will be an AC voltage whose amplitude corresponds to the amount of energy contained within the filter's bandwidth. The outputs of each of these filters can then be detected and displayed (either together on a CRT or with individual displays) to produce the frequency domain information (Figure 8-3). This type of instrument is called a BANK-OF-FIL-TERS ANALYZER.

8.2.1 Frequency Resolution

The frequency resolution of the bank-of-filters analyzer is determined by the bandwidth of the individual filters. Figure 8-2 shows that both filter 1 and filter 2 will detect unique spectral lines. The frequency of each of those lines is known only to the

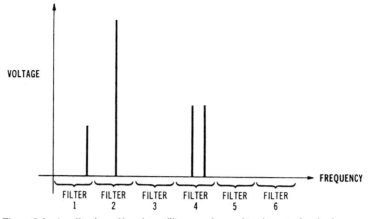

Figure 8-2 A collection of bandpass filters can be used to characterize the frequency content of a signal. Each filter measures the energy in a small frequency band.

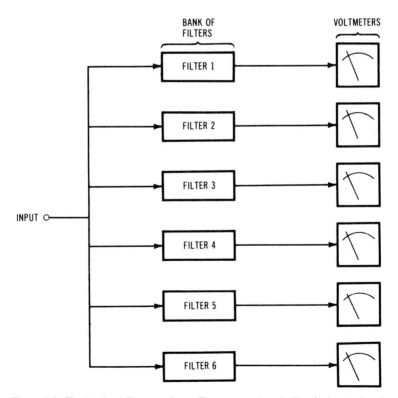

Figure 8-3 The bank-of-filters analyzer. The output of each filter is detected and displayed to provide the frequency domain characteristics of a signal.

extent that they are within the frequency range of their respective filter. If the filter is 100 Hz wide, then the frequency of the spectral line has an uncertainty of 100 Hz. Another situation is shown in filter 4 where two spectral lines are shown within the same filter's frequency range. These two spectral lines are not detected as being distinct frequency components. Instead, the filter will measure the amount of energy present within its frequency range without regard to how many spectral lines produced that energy. Thus, the ability to resolve two closely spaced spectral lines also depends on the width of the filters.

The bank-of-filters analyzer is simple in concept and results in an instrument capable of quickly tracking changes in the signal's spectral content. The major disadvantage of this approach is that a very large number of narrow filters is needed for most applications. The number of filters required increases with the frequency range and frequency resolution of the analyzer. For example, an analyzer designed to cover 0 to 1 MHz with 1-kHz-wide filters would require 1000 filters. This limitation prevents the technique from being used for wide-bandwidth instruments. However, this technique is used in low-frequency applications (such as audio and sound-level measurements) where a more reasonable number of filters is sufficient.

Although the bank-of-filters technique is not often used in general electronic measurement, it does provide a good conceptual basis for understanding the instrumentation techniques that are commonly used.

8.3 FFT SPECTRUM ANALYZERS

Another technique used to measure the frequency spectrum of a signal is the FAST FOURIER TRANSFORM (FFT). The FFT is a very efficient mathematical technique for computing the spectrum of a waveform from its time domain representation. First, the analog waveform must be turned into digital form by means of an analog-to-digital converter. Then, the data is fed into a microprocessor which computes the frequency spectrum. The spectrum is usually displayed on a CRT or other display device.

8.3.1 Sampling

The waveform must be turned into digital form before the FFT computation can be applied to it. This is accomplished by sampling the waveform at regular intervals and converting the voltage at each point to a digital number. Figure 8-4 shows a sine wave and the sampled data points taken from it. Each one of these sample points is converted into a digital number. Notice that since the points are fairly closely spaced on the sine wave, the waveform can easily be reconstructed by filling in between the sample points. The rate at which the sampling occurs is called the SAMPLE RATE and is usually expressed in units of samples per second or hertz.

Care must be taken to make sure that enough samples are taken on each cycle of the waveform so that the frequency spectrum of the waveform can be extracted from the sample points. How fast do we need to sample a given signal? The SAMPLING THEOREM states that all the information in a signal is preserved if it is

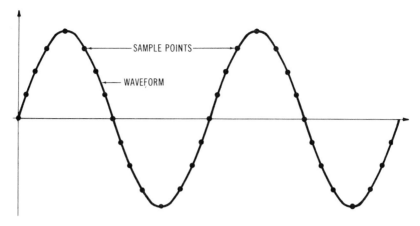

Figure 8-4 The input waveform must be sampled for the FFT to be computed.

sampled at a rate that is over twice the highest frequency present. For a spectrum analyzer having bandwidth BW (BW = the highest frequency measured) and sample rate f_s, the following equation must be true:

$$f_s > 2 \, BW$$

This is the theoretical limit. Practical considerations require the sample rate to be even higher.

Other problems can occur when waveforms are sampled. Figure 8-5 shows two waveforms and a set of sample points. Notice that the sample points fall on both these waveforms. Stated differently, with the sample rate shown, both waveforms would produce the exact same sample points. Therefore, the two waveforms cannot be distinguished after sampling has occurred. This phenomenon is called ALIASING, since

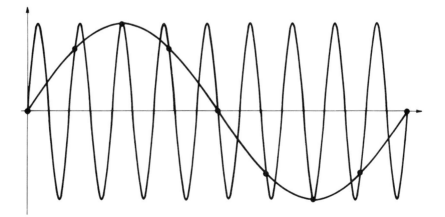

Figure 8-5 Aliasing can occur if precautions are not taken. Shown here are two different waveforms that will result in the same set of sample points. Therefore, after sampling the two waveforms are indistinguishable.

one waveform acts as the alias (or impostor) of the other. This aliasing is highly undesirable in an instrument that is intended to measure the frequency content of a signal.

A simple solution exists for the aliasing problem. Notice that in Figure 8-5, one waveform has a much higher frequency than the other one. Consistent with the sampling theorem, the spectrum analyzer is designed to measure frequencies below about half of the sample rate. If a low-pass filter is added before the ADC, then the undesirable high-frequency signal is removed. The resulting block diagram for the FFT spectrum analyzer is shown in Figure 8-6. The signal being measured passes through the input amplifier and is low pass filtered to remove any potential alias signals. The waveform is then sampled and converted to digital numbers by the ADC. The microprocessor receives the digital information and performs the FFT calculation. This results in the frequency spectrum, which is then displayed.

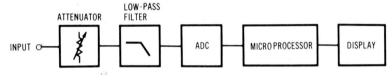

Figure 8-6 Simplified block diagram of an FFT spectrum analyzer. The low-pass filter is included to prevent aliasing.

The FFT computation is a very involved and complex set of mathematical operations. A stream of digital samples from the time domain is collected and processed as one array of data. The output of the FFT is also an array of data, but in this case it is frequency domain data. The FFT operates as a time domain–to–frequency domain converter, taking a slice of time domain data and converting it into its corresponding frequency spectrum (Figure 8-7). The output of the FFT computation is equivalent to the output of a bank-of-filters analyzer. Each frequency domain data point corresponds to the output of one of the filters in the bank-of-filters analyzer. The FFT does not require a large number of filters since the same effect is accomplished mathematically.

A typical FFT analyzer has about 400 such data points, which means that 400 filters would be required in a bank-of-filters analyzer to achieve the same frequency resolution. Most FFT analyzers have the ability to vary the frequency span and hence

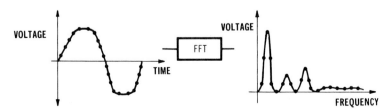

Figure 8-7 The FFT computation takes a slice of the sampled time domain data and computes the frequency domain from it.

their frequency resolution (with a constant number of frequency data points). Thus, the bank-of-filters analyzer would need considerably more than 400 filters to achieve the same overall performance as an FFT analyzer. Very narrow filter bandwidths are difficult to build using the bank-of-filters approach. However, the FFT computation can produce the equivalent of extremely narrow filters, with bandwidths much smaller than 1 Hz.

The major limitation of the FFT analyzer is the ADC. Existing ADC technology (at a reasonable cost) tends to limit the bandwidth of the typical FFT spectrum analyzer to around 100 kHz. Some recent high-performance analyzers have extended the bandwidth to 10 MHz. ADC technology will continue to improve, providing faster sample rates and allowing wider bandwidth FFT analyzers to be developed. Another limitation is the computing power of the microprocessor or computer used to compute the FFT. The amount of time it takes to compute the FFT may be significant, limiting the speed with which the spectrum can be displayed. This, too, will undoubtedly improve with time. The specifications of a typical FFT spectrum analyzer are shown in Table 8-1. Figure 8-8 shows a spectrum analyzer that uses FFT techniques to characterize signals in the frequency domain.

TABLE 8-1 SPECIFICATIONS OF A TYPICAL FFT SPECTRUM ANALYZER.

Frequency range: 0.000125 Hz to 100 kHz
Measurement range: + 27 to − 120 dBV
Dynamic range: 80 dB
Amplitude accuracy: ± 0.15 dB

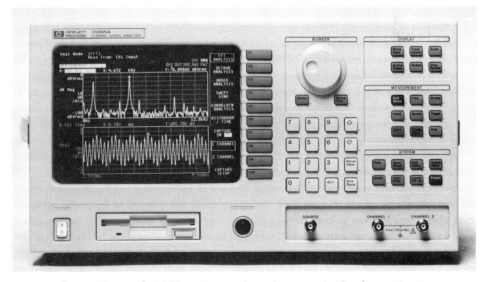

Figure 8-8 A typical FFT spectrum analyzer that uses a fast Fourier transform to compute the spectrum of a signal. (Photo courtesy of Hewlett-Packard Company.)

8.4 WAVEMETERS

The bank-of-filters spectrum analyzer uses many filters, each one tuned to a different frequency, to measure the frequency components present in a waveform. The WAVE-METER accomplishes this by having one filter whose frequency is tunable over the frequency range of interest (Figure 8-9). The tunable filter is tuned to the frequency of interest, and the output level of the filter is measured and displayed by a meter. The meter can display only one frequency component at a time, namely, the frequency to which the filter is tuned. This is obviously a disadvantage compared to the FFT and bank-of-filters spectrum analyzers, but it is quite acceptable for many applications.

Generally, it is very difficult to design and build filters that can be tuned over a wide frequency range while maintaining a constant filter shape. Thus, the frequency tunable filter is not usually implemented this way. (One notable exception is at microwave frequencies where certain technologies allow a variable frequency filter to be built.)

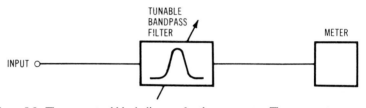

Figure 8-9 The conceptual block diagram for the wavemeter. The wavemeter uses a frequency tunable filter to measure the signal level at a single frequency.

8.4.1 The Practical Wavemeter

Instead of tuning the filter over a frequency range, it is easier to leave the filter at a fixed frequency and move the signal in frequency. This technique is shown in Figure 8-10. Starting from the right side of the figure, the filter is designed to measure the level at the intermediate frequency (IF). The mixer and local oscillator (LO) shift the input signal in frequency to the intermediate frequency. Thus, the filter can be at a fixed frequency at the expense of shifting the input signal in frequency.

A mixer takes the input frequency (f_{IN}) and the local oscillator frequency (f_{LO}) and produces the sum and difference frequencies:

$$f_{LO} - f_{IN}$$

$$f_{LO} + f_{IN}$$

So for every input frequency, there are two frequencies at the output of the mixer.[1] One of these frequencies is ignored by the IF filter, and the other is passed

---

[1]Actually there will be many different frequencies present at the output, but for the purposes of understanding the block diagram, only two are significant.

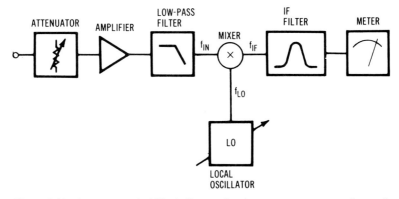

Figure 8-10 A more practical block diagram for the wavemeter uses a mixer and local oscillator to shift the input signal in frequency. The intermediate frequency filter is tuned to the output of the mixer.

through and measured (assuming that the wavemeter is tuned to that frequency). The local oscillator's frequency is variable and is adjusted by the user to cause the mixing process to produce the desired frequency at the IF filter. The function of the low-pass filter will be discussed later.

This method of frequency translation is known as the superheterodyne technique and is the same technique used in virtually every modern radio receiver. A numerical example will help to clarify this mixing process. Figure 8-11 shows a wavemeter designed to measure from 0 to 10 MHz. The LO operates from 20 to 30 MHz, and the IF filter is tuned to 20 MHz. Suppose the wavemeter was tuned to measure the level at 5 MHz. The LO is tuned to 25 MHz, resulting in 25 ± 5 MHz (equals 20 MHz and 30 MHz) out of the mixer. The 20-MHz signal falls directly on the IF filter and is measured, while the 30-MHz signal is ignored. If the wavemeter were tuned to 6 MHz, then the LO would be tuned to 26 MHz. The mixer output

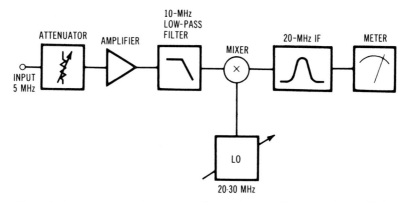

Figure 8-11 An example of a wavemeter block diagram with frequencies specified.

frequencies would be 20 MHz and 32 MHz. The instrument controls are, of course, designed so that the LO frequency adjustment indicates the wavemeter frequency and not the frequency of the LO.

So what about the low-pass filter (also known as the IMAGE FILTER)? Again referring to Figure 8-11, assume that the wavemeter is tuned to 5 MHz (with the LO at 25 MHz). If a 45-MHz signal were present at the input to the mixer, it would mix with the 25-MHz LO frequency and produce sum and difference frequencies (20 and 70 MHz). The 70-MHz signal would not cause any problems, but the 20-MHz signal would fall directly on the IF. Therefore, the wavemeter would not be able to distinguish between a 5-MHz signal and a 45-MHz signal. The undesired signal at 45 MHz is known as an IMAGE and is eliminated by the image filter. Image signals occur at frequencies equal to the LO frequency plus the IF frequency (for the block diagram shown in Figure 8-11). Or equivalently, the image is at the input frequency plus two times the intermediate frequency.

As a practical matter, high-performance wavemeters are more complex than the single IF block diagram shown. Often several sets of intermediate frequency filters and mixers are used to implement a wavemeter. Conceptually, these block diagrams use the same frequency translation techniques to move the signal past the filter to eliminate the need for a variable-frequency filter. Although practical wavemeters usually do not use variable-frequency filters, the concept is still a valuable tool for understanding the function performed by a wavemeter. In fact, the super-heterodyne block diagram produces the same effect as the variable-frequency filter approach.

Suppose the frequency components of a 1-MHz square wave are to be measured. Figure 8-12 shows the time domain representation and the frequency domain points measured by a wavemeter. Tuning the wavemeter to 1 MHz allows the amplitude of the fundamental to be measured, while ignoring the harmonics. Similarly, when the wavemeter is tuned to 3 MHz and then 5 MHz, the 3rd and 5th harmonics, respectively, are measured (while ignoring the other frequency components). Other frequencies can be measured as needed, but the wavemeter is a "one frequency at a time" kind of instrument. Note that a simple voltmeter cannot make this measurement, since it is inherently a broadband instrument without controlled frequency selectivity. Performing this type of measurement

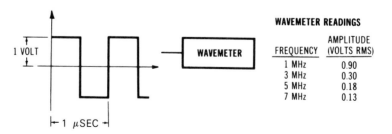

	WAVEMETER READINGS	
		AMPLITUDE
FREQUENCY		(VOLTS RMS)
1 MHz		0.90
3 MHz		0.30
5 MHz		0.18
7 MHz		0.13

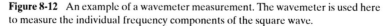

Figure 8-12 An example of a wavemeter measurement. The wavemeter is used here to measure the individual frequency components of the square wave.

should convince the reader that periodic signals really do have harmonics in the frequency domain.

8.5 RESOLUTION BANDWIDTH

The bandwidth of the IF filter determines the frequency resolution of the instrument (similar to the bank-of-filters spectrum analyzer). Thus, the bandwidth of the IF filter is often referred to the RESOLUTION BANDWIDTH. If more than one IF is used in an instrument, the narrowest one dominates and is considered the resolution bandwidth. (Don't confuse the resolution or IF bandwidth with the bandwidth of the instrument, which refers to the overall frequency range of the instrument.) Most wavemeters have selectable resolution bandwidths to allow measurement flexibility.

The choice of resolution bandwidth depends on several factors. Filters take some time to settle. That is, when a signal first appears at the input of a filter, it will take a while before the signal appears at the output. In addition, the output of the filter will take some time to settle to the correct value, so that it can be measured. The narrower the filter bandwidth, the longer the settling time. In many cases, this time is so small that it is inconsequential. But if very narrow bandwidths are used, the settling time can become significant.

Narrow bandwidths also increase the difficulty of tuning to a particular signal, particularly if the signal frequency is unknown or varies. The wavemeter must be tuned exactly to the signal frequency; otherwise, the signal will fall outside the resolution bandwidth. If the signal or wavemeter frequency drifts, then maintaining the same frequency is difficult. Using a wider resolution bandwidth makes the frequency difference between the signal and wavemeter less critical. Compounding the problem is the fact that narrower bandwidths take longer to settle. For example, a 10-Hz-wide filter might take on the order of a second to settle to an accurate value. If an unknown frequency is being measured, it takes several (if not many) adjustments to tune the wavemeter to the signal frequency. So as narrower bandwidths are used, the settling and adjustment times increase while the required accuracy of the frequency adjustment also increases.

The choice of resolution or IF bandwidth will depend on the signal being measured. Figure 8-13 shows two signals very close together in frequency. If the two spectral lines are to be measured individually, then a narrow bandwidth is required (as shown). If a wider bandwidth is used, then the energy of both signals will be included in the measurement. (This could be desirable in some cases.)

There is always some amount of noise present in a measurement. Often it is so small that it can be ignored, but for low-level measurements, noise will become important. Noise is generally broadband in nature; that is, it exists at a broad range of frequencies in the frequency domain. If the noise is included in the measurement, the measured value will be in error (too large) depending on the noise level. Figure 8-14 shows a signal being measured in the presence of noise. With a wide bandwidth, more noise is included in the measurement. With a narrow bandwidth, very little noise enters the resolution bandwidth filter, and the measurement is more accurate. The

FREQUENCY RESOLUTION OF FILTER

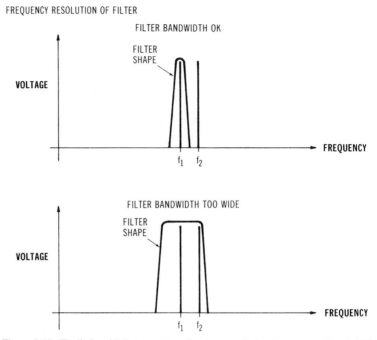

Figure 8-13 To distinguish between two closely spaced signals, a narrow bandwidth is required. If a wide bandwidth is used, then both signals will be included in the measurement.

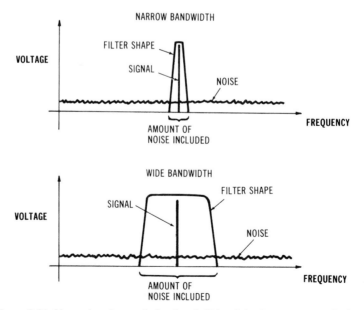

Figure 8-14 Narrowing the resolution bandwidth of the instrument results in a smaller amount of noise corrupting the measurement.

effect of noise on a voltage measurement is proportional to the square root of the resolution bandwidth.

8.6 NARROWBAND VERSUS BROADBAND MEASUREMENTS

The bank-of-filters analyzer, the FFT analyzer, and the wavemeter are all examples of narrowband measurements. Although the first two instruments measure a range of frequencies simultaneously, the techniques they use are the same as a narrowband measurement. The wavemeter is more obviously narrowband because it acts like a variable frequency filter tuned to the frequency of interest. On the other hand, instruments such as voltmeters and oscilloscopes are broadband. That is, they do not have the ability to look at frequencies selectively, but instead must measure across their entire frequency range at all times.

As an example, consider the frequency domain representation of the sine wave shown in Figure 8-15a. Ideally, the sine wave is a single, infinitely thin spectral line. In reality, some or all of the imperfections shown in Figure 8-15b (harmonics, spurious responses, and noise) may be present. A broadband measuring instrument such as a voltmeter would include the fundamental, harmonics, spurious responses, and noise in the measurement. This may be desirable if the goal of the measurement is to determine the total signal level present across a wide bandwidth. In many other cases, it is desirable to ignore the imperfections, especially the noise and spurious responses that are not part of the signal. (Is the harmonic part of the signal? Sometimes.) A narrowband measuring instrument such as a wavemeter will filter out all but the desired frequency components, resulting in a measurement that includes only the fundamental.

But suppose the harmonics or the spurious responses were to be measured. The voltmeter is useless, but the wavemeter could be tuned to the proper frequency to make the measurement. (The bank-of-filters and FFT spectrum analyzers could also be used to make the measurement. In fact, they would be more convenient since they measure the entire frequency range at one time.) What about the noise level: Could it be measured in the presence of a large signal? Again, the broadband voltmeter would be useless since it includes everything in the measurement. A narrowband instrument, however, could be tuned to the frequency of interest and (assuming no spectral lines happened to be at that frequency) could measure the noise level there. The bandwidth of the measurement would have to be taken into account since the resolution bandwidth will affect how much noise is measured.

8.7 SWEPT-SPECTRUM ANALYZERS

The wavemeter block diagram can be improved one step farther. The wavemeter measures at only one frequency at a time, but if the wavemeter could be tuned or swept across its frequency range, then the entire spectrum of a signal could be automatically characterized. The SWEPT-SPECTRUM ANALYZER is essentially a wavemeter that automatically sweeps in frequency and displays the results. Like

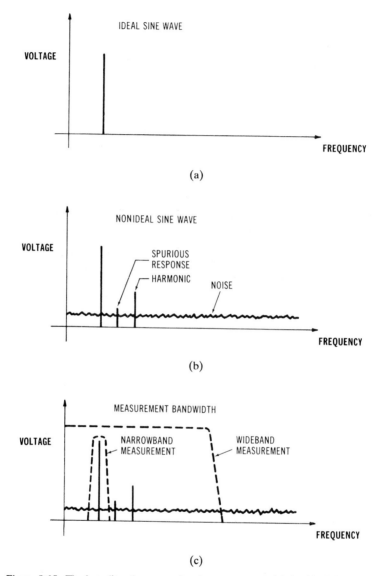

Figure 8-15 The benefits of a narrowband measurement. (a) An ideal sine wave is a single spectral line. (b) The sine wave may be accompanied by a variety of imperfections. (c) The broadband measurement includes everything in the measurement. A narrowband measurement can select the desired frequency component.

the bank-of-filters and the FFT spectrum analyzers, the result is a frequency domain display of the signal (with voltage on the vertical axis and frequency on the horizontal axis). Figure 8-16 shows the concept of sweeping the wavemeter.

Figure 8-17 shows a simplified block diagram of a swept-spectrum analyzer. A voltage controlled oscillator is used as the local oscillator to allow the frequency

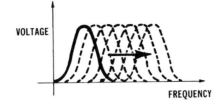

Figure 8-16 A swept-spectrum analyzer can be thought of as a wavemeter which is swept in frequency. The result is a filter that automatically sweeps in frequency, characterizing the signal at its input.

of the analyzer to be changed easily. A ramp generator drives the tuning voltage of the VCO, causing it to sweep up in frequency repetitively. The same ramp voltage drives the horizontal axis of the display. The image filter, mixer, and IF filter all function the same as they did with the wavemeter. The output of the IF filter, however, goes to the DETECTOR, which detects the sine wave, producing a DC level that is proportional to the level present in the IF filter. This signal drives the vertical axis of the display, resulting in the IF level being painted across the display while the VCO sweeps. The result is an amplitude versus frequency display of the signal.

The operation of the analyzer has been described in an analog sense. Digital technology has replaced some of the analog circuits. In particular, the output of the detector is typically measured by an ADC which is connected to a microprocessor. In some cases, the ADC has been moved into the IF section, and the resolution bandwidth filtering and detection are done digitally. The microprocessor receives the data in digital form and processes it to the display. The input amplifier, image filter, mixer, and VCO are all implemented using analog technology. This allows the analyzer to operate at very high frequencies (even microwave and above) which would not be possible with digital techniques. As digital technology

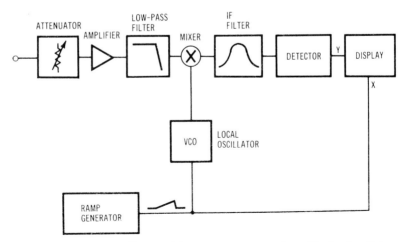

Figure 8-17 The block diagram of the swept-spectrum analyzer. The local oscillator is swept in frequency while the output of the intermediate frequency filter is displayed on the screen.

improves, its use will migrate forward in the block diagram toward the input amplifier.

8.7.1 Sweep Limitations

The user normally has control over the SWEEP TIME (the elapsed time of each sweep, sometimes called SCAN TIME), the frequency range over which the analyzer sweeps, and the resolution bandwidth. The analyzer cannot be swept arbitrarily fast while maintaining its specified accuracy, but will have a sweep rate limitation depending on the resolution bandwidth chosen. The sweep rate (expressed in hertz per second) is not usually chosen directly by the user, but is determined by the frequency range swept divided by the sweep time.

The limitation on sweep rate comes from the settling or response time of the resolution bandwidth filter discussed earlier. If an analyzer is swept very quickly, the filter does not have time to respond, and the measurement is inaccurate. Under such conditions, the analyzer display tends to have a "smeared" look to it, with spectral lines being wider than normal and shifted to the right. In general, the maximum sweep rate of an analyzer is given by

$$\text{sweep rate} = \frac{\text{BW}^2}{k}$$

where BW is the bandwidth of the resolution filter and k is a factor depending on the resolution bandwidth filter shape (typically k is equal to 2). Notice that the sweep rate is proportional to the bandwidth squared. There may be other limitations on sweep rate in the instrument, such as the speed at which the local oscillator can sweep.

Fortunately, most instrument manufacturers have designed in mechanisms that unburden the user from having to calculate the sweep rate. In older instruments, this is usually accomplished by some sort of mechanical interlock and/or a warning light which tells the user that the measurement may be uncalibrated. In later-model instruments, a microprocessor usually chooses the fastest accurate sweep time. In both cases, an informed user is protected from making an erroneous measurement. At the same time, the user is given the option of overriding the built-in sweep rate protection. There may be valid reasons for doing this, but the user must proceed at his or her own risk (of making an inaccurate measurement).

Since the sweep rate is proportional to the square of the resolution bandwidth, decreasing the resolution bandwidth dramatically decreases the maximum sweep rate. For a given swept-frequency range, the sweep time is inversely proportional to the square of the resolution bandwidth. Thus, the user will find that the sweep time (and, therefore, the measurement time) must be significantly increased when narrow resolution bandwidths are used. This illustrates one of the advantages of the FFT analyzer over the swept analyzer. FFT techniques result in much narrower resolution bandwidths for a given measurement time, or they give the same resolution bandwidth much faster.

8.7.2 Dynamic Range

The DYNAMIC RANGE of a spectrum analyzer is the difference between the largest signal and the smallest signal that can reliably be measured AT THE SAME TIME (Figure 8-18). (Compare this to MEASUREMENT RANGE which is the difference between the largest signal that can be measured and the smallest signal that can be measured, but not simultaneously.) The smallest signal that can be measured is limited by noise, distortion, and spurious responses present in the analyzer. Noise is present in all electronic circuits, although it is often so small compared to the signals present that it is inconsequential. For low-level spectrum analyzer measurements, even a small amount of noise is significant. When a signal falls below the noise level present in the instrument, that signal can no longer be reliably measured. Spectrum analyzers also produce a certain amount of distortion in their measurement circuits. This means that a perfectly clean signal may be displayed as having some small harmonics or other spectral impurities (due to the distortion in the instrument). So when a signal falls below these distortion products present in the analyzer, the user cannot distinguish between a valid signal and an erroneous one. In addition, some residual spurious responses may be present in the instrument. Distortion products will disappear when the input signal is removed, but residual responses will be present with or without an input signal. All these imperfections in the instrument are specified by the manufacturer, either separately or in terms of dynamic range.

8.7.3 Effect of Resolution Bandwidth

As previously mentioned, using a narrower resolution bandwidth causes a smaller amount of noise to be present in the measurement. The amount of noise in the circuit

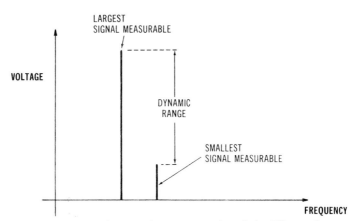

Figure 8-18   The dynamic range of a spectrum analyzer is the difference between the largest signal and the smallest signal measurable at the same time. The smallest signal measurable is limited by noise, distortion, and residual responses.

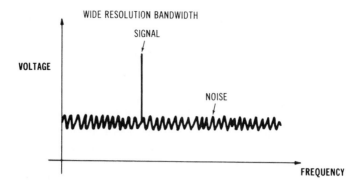

Figure 8-19 The effect of changing the resolution bandwidth. A wider resolution bandwidth has a higher noise level on the display.

remains the same, but a smaller amount of it is present at the detector (Figure 8-19a shows a typical spectrum measurement having a signal present above the broadband noise "floor." Note that the noise is generally present at all frequencies and may be present in the signal being measured or may be the internal noise of the analyzer. If the resolution bandwidth is narrower, the noise floor drops on the display (Figure 8-19b). Again, this is because the IF filter of the analyzer has been made narrower in bandwidth, which lets in less noise. A factor of 2 change in resolution bandwidth causes a 3-dB change in the noise level. As the measured noise level drops, smaller signals that were previously obscured by the noise can be measured. This has the effect of increasing the dynamic range of the measurement at the expense of increased sweep time.

8.7.4 Effect of Video Bandwidth

Most analyzers include another type of filtering after the detector called VIDEO FILTERING. This filter also affects the noise on the display, but in a different manner than the resolution bandwidth. Figure 8-20 shows the positioning of the

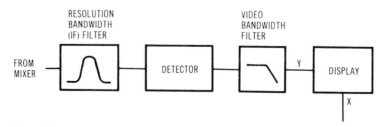

Figure 8-20 A portion of the swept analyzer block diagram showing the position of the video bandwidth filter.

video filter in the swept-spectrum analyzer block diagram. Note that the resolution filter is in front of the detector (called predetection filtering) and the video filter is after the detector (called postdetection filtering). The effect of video filtering is shown in Figure 8-21. Figure 8-21a shows a signal being measured in the presence of noise with a wide video filter. Figure 8-21b shows the same signal and noise with a narrow video filter. The average level of the noise remains the same, but the variation in the noise is reduced. The effect on the analyzer's display is that the noise floor compresses into a thinner trace, while the position of the trace remains the same. (Compare this with the effect of the resolution bandwidth, which reduces the level of the noise.)

8.7.5 Tracking Generator

Many swept-spectrum analyzers include a sine wave output which tracks the analyzer's input frequency as it sweeps. Since this sine wave is always at the same frequency that the analyzer is measuring, it allows convenient frequency response

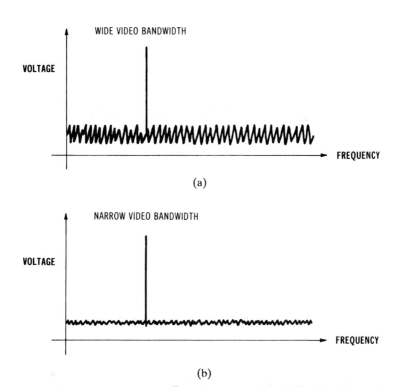

Figure 8-21 The effect of changing the video bandwidth. (a) Signal in the presence of noise with a wide video bandwidth. (b) The same signal and noise with a narrow video bandwidth. The average noise level remains the same, but the variation in the noise is reduced.

characterization of circuits. This TRACKING GENERATOR output is connected to the input of the circuit being tested, and the output of the circuit is connected to the input of the analyzer (Figure 8-22). Recall from Chapter 5 that characterizing the frequency response of a circuit (such as a filter or amplifier) can be done on a point-by-point basis using a source and a scope. A spectrum analyzer with a tracking generator can accomplish the same thing, but much more quickly. The resulting frequency response is plotted on the spectrum analyzer display.

8.7.6 Advanced Features

Modern spectrum analyzers are often microprocessor controlled, which allows more advanced features to be implemented. A marker (or cursor) is supplied which can be located at any point on the displayed trace, with the frequency and amplitude values at that point displayed digitally on the display. A marker reference or offset can be specified so that the marker reads relative to the reference point. The marker can usually be configured to display whatever units are convenient to the user (dBm, dBV, volts, etc.). Other marker features allow the user to assign the current marker frequency to the center frequency and the current marker value to the full-scale value (reference level).

Digital storage of the displayed trace keeps the display from flickering at the sweep rate. Additional trace storage may be provided, as well as the ability to display the difference between the current and stored traces. Other features provided include automatic noise measurement, a built-in frequency counter, and a manual sweep (not a sweep at all, but a single-frequency measurement).

A typical portable spectrum analyzer is shown in Figure 8-23. The specifications of a typical swept-spectrum analyzer are shown in Table 8-2. The most notable difference between the FFT analyzer and the swept analyzer is the frequency range. It should be pointed out that the typical specifications shown are for a particular radio-frequency analyzer—commercially available microwave spectrum analyzers have frequency ranges that extend beyond 40 GHz.

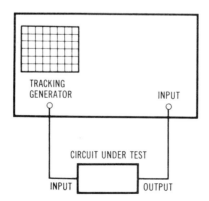

Figure 8-22 A spectrum analyzer having a tracking generator output can be used to measure the frequency response of a circuit.

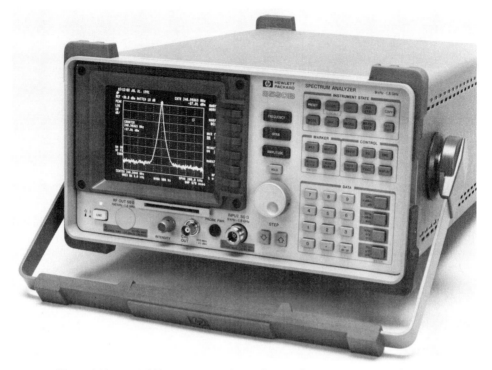

Figure 8-23 A 1.8-GHz spectrum analyzer with tracking generator output. (Photo courtesy of Hewlett-Packard Company.)

TABLE 8-2 SPECIFICATIONS OF A PORTABLE SWEPT-SPECTRUM ANALYZER.

Frequency range: 9 kHz to 1.8 GHz
Measurement range: +30 to −115 dBm
Dynamic range: 70 dB
Amplitude accuracy: ± 1.0 dB

8.8 SPECTRUM ANALYZER MEASUREMENTS

The spectrum analyzer (regardless of the implementation) gives the instrument user a means of viewing the frequency domain just as the oscilloscope provides a picture of the time domain. Accordingly, a variety of measurements can be made using the spectrum analyzer that simply involve interpreting the frequency domain display of a signal. The simplest type of spectrum analyzer measurement is just measuring the amplitude and frequency of a single sine wave. Although analyzers are well suited for such measurements, the real power of the spectrum analyzer is utilized when more complex spectra are measured.

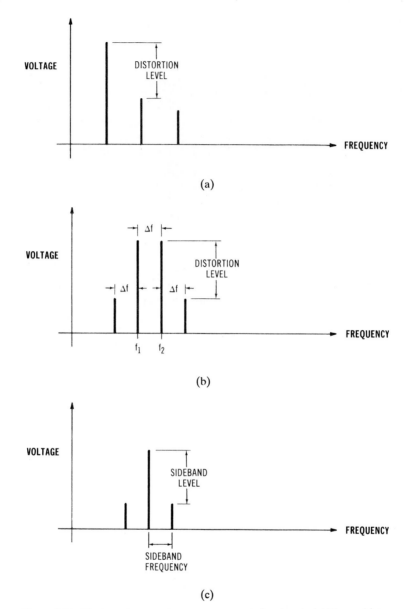

Figure 8-24 Most spectrum analyzer measurements involve straightforward inter-pretation of the frequency domain display provided by the analyzer. (a) Harmonic distortion measurement. (b) Intermodulation distortion measurement. (c) Measure-ment of modulation sidebands.

HARMONIC DISTORTION is present in most real-world sine waves as shown in Figure 8-24a. Ideally, a sine wave would be a single spectral line at the fundamental frequency. With harmonic distortion present, some of the waveform's energy appears at the harmonic frequencies. Note that the spectral lines are at multiples of the fundamental frequency. It is interesting to note that a good spectrum analyzer is capable of measuring distortion products that are 80 dB smaller than the fundamental (0.01 percent distortion). That is equivalent to a factor of 10,000 in voltage. Suppose, for example, that amount of distortion is present in a 1-volt RMS sine wave. The corresponding harmonic distortion level is $1/10,000 = 100 \,\mu V$. Harmonic distortion that small would not be detectable by the most sensitive oscilloscope (because the scope must display the rather large fundamental waveform simultaneously). The best oscilloscope displays are capable of resolving distortion in the range of a few percent for a trained operator. Thus, the spectrum analyzer is far superior to an oscilloscope for making distortion measurements. Harmonic distortion is generally measured with a spectrum analyzer in decibels relative to the fundamental, but can also be expressed in percent.

INTERMODULATION DISTORTION occurs when two sine waves are present at the same time in a circuit that exhibits distortion. Ideally, the addition of two sine waves simply results in two unique spectral lines. If distortion is present in the system, the two signals will intermodulate. In addition to producing the harmonics of both waveforms, this causes several types of sum and difference frequencies to be generated. For instance, as shown in Figure 8-24b, it is common for spectral lines to appear at the original frequencies plus or minus the difference between the two original frequencies.

MODULATION SIDEBANDS may appear due to intentional or unintentional amplitude modulation (AM), frequency modulation (FM), or phase modulation (PM). Modulation is intended to be on many communication signals, including standard AM and FM radio broadcast signals. In other cases, the modulation may be a by-product of some other circuit operation. In either case, it is important to be able to characterize the signal accurately. The spectrum analyzer inherently measures the absolute sideband level, which may be expressed in decibels relative to the carrier frequency. Depending on the type of modulation, this relative sideband level can be related back to percent modulation, frequency deviation, and so on.[2]

8.9   SPECTRUM ANALYZER INPUTS

At frequencies below 50 MHz, spectrum analyzers are usually provided with both high-impedance (1-MΩ oscilloscope-type inputs) and 50-Ω inputs. As the upper limit of the frequency range increases, the high-impedance input becomes impractical due to the effect of the input capacitance and only the 50-Ω input is included. Although not as common as 50-Ω inputs, 75-Ω inputs are also found on some spectrum analyzers. The high-frequency swept analyzer is usually fundamentally a low-impedance instrument for noise and frequency response reasons. The high-impedance input usually has degraded specifications associated with it, and the 50- (or 75-) Ω input is the "quality" input. Lower-frequency FFT instruments usually are supplied with only high-impedance inputs.

[2]For more information on spectrum analyzer measurements, see Witte (1991).

8.10 GAIN/PHASE METERS

As shown earlier, a source and oscilloscope can be used to measure the amplitude and phase relationships of two signals (usually the input and output of a circuit). Other instruments have been developed to provide a more convenient means of making such a measurement. A GAIN/PHASE METER is a two-channel instrument which automatically compares the amplitude levels and phases of two signals and displays the results. The most common use of the gain/phase meter is in measuring the gain and phase relationships of a circuit such as a filter or amplifier (Figure 8-25). The instrument may also display the absolute level of each channel.

The block diagram of a gain/phase meter is shown in Figure 8-26. Each input (A and B) drives a (magnitude) detector, which produces a DC voltage proportional to the signal level. The ratio of these two DC voltages is determined by another circuit which drives the gain display. The resulting gain display is simply a comparison of the A and B input levels. If the A and B inputs are connected to the input and output, respectively, of a circuit, the gain display shows the gain of the circuit under test.

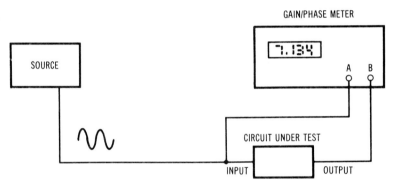

Figure 8-25 The gain/phase meter is most often used to measure the magnitude (gain) and phase characteristics of a circuit (requires external source).

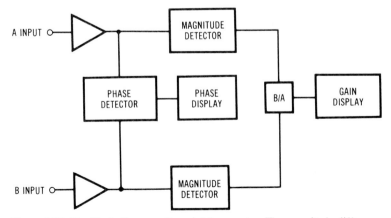

Figure 8-26 The block diagram of a gain/phase meter. The magnitude difference (gain) and phase difference between the two inputs is measured and displayed.

Figure 8-27   A gain/phase meter is used to measure the amplitude and phase difference between two signals. (Photo courtesy of Hewlett-Packard Company.)

Similarly, the phase detector compares the phases of the two inputs, generating a DC voltage which corresponds to the phase shift between the two inputs. This voltage drives the phase display. The gain and phase displays may be implemented using analog or digital technologies. A gain/phase meter is shown in Figure 8-27, and the specifications for a typical gain/phase meter are listed in Table 8-3.

TABLE 8-3   SPECIFICATIONS OF A TYPICAL GAIN/PHASE METER.

Frequency range: 1 Hz to 13 MHz
Measurement range: +26 to −74 dBV
Dynamic range: 80 dB
Amplitude accuracy: ±1 dB
Phase accuracy: ±2 degrees

8.11 NETWORK ANALYZERS

Although the gain/phase meter is a convenient way to make single-point gain and phase measurements, many single-point measurements are required to characterize fully a circuit over a wide frequency range. In addition, the gain/phase meter is usually a broadband measuring instrument, which means that noise present in the circuit will limit the dynamic range. A NETWORK ANALYZER is essentially a gain/phase meter that automatically measures the gain and phase of a circuit over a selectable frequency range. A network analyzer may have a sine wave source built in, or the analyzer may control an external source. The source sweeps in frequency while the input of the analyzer measures the gain and phase through the circuit (Figure 8-28). Network analyzers typically have more than one input channel for increased measurement flexibility. Figure 8-29 shows a two-channel network analyzer using a power splitter to supply the source signal simultaneously to the device under test and the A channel of the analyzer. The analyzer is then set to measure the ratio of channel B and channel A (B/A). Variations in the source level will be measured by channel A and effectively removed from the measurement. Without ratioing, these source variations would cause errors in the measurement. Another possible use for multiple

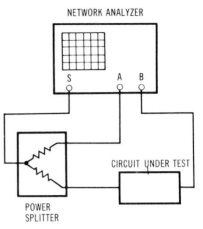

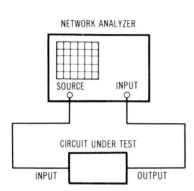

Figure 8-28 A network analyzer provides the stimulus to the device under test and measures the circuit's response.

Figure 8-29 A typical measurement connection using a two-channel analyzer and a resistive power splitter. The source supplies a sine wave to both the A input and the circuit under test. The analyzer is configured to measure the ratio of channel B and channel A.

channels is the simultaneous display and comparison of the characteristics of two different circuits.

Network analyzers are usually designed using techniques very similar to a swept-spectrum analyzer. The use of narrowband IF filters allows a network analyzer to achieve a dynamic range in excess of 100 dB. The conceptual block diagram of a network analyzer is shown in Figure 8-30. The network analyzer block diagram is similar to the spectrum analyzer's, although there are some practical differences in how they are actually implemented. A network analyzer is always assumed to be measuring a known frequency generated by the analyzer's source. A spectrum analyzer, on the other hand, must be capable of measuring unknown and arbitrary frequencies. Thus, the network analyzer may not include the same level of image rejection that a spectrum analyzer does. A true network analyzer must include a means of measuring phase, while this is not necessary on a spectrum analyzer. Some instruments include both spectrum and network analyzer functions for maximum flexibility. A typical network analyzer is shown in Figure 8-31. The specifications of a typical network analyzer are shown in Table 8-4.

8.11.1 Network Analyzer Measurements

The display of a network analyzer provides a plot of the magnitude and/or phase response of a two-port circuit as a function of frequency. The most common format for the display is gain (or loss) in dB versus logarithmic frequency. Other types of displays are also usually provided such as polar (Smith chart) plots and group delay versus frequency. Some examples of network analyzer plots are shown in Figure 8-32.

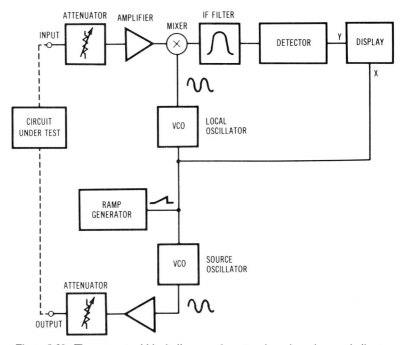

Figure 8-30   The conceptual block diagram of a network analyzer is very similar to a spectrum analyzer block diagram except for the addition of a source output and its associated circuitry.

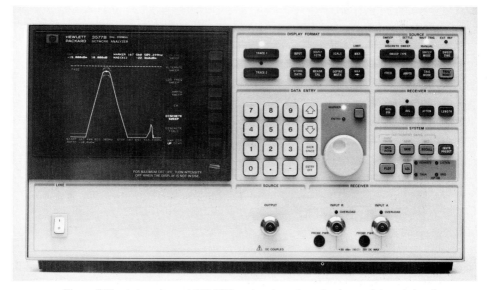

Figure 8-31   A two-channel 200-MHz network analyzer having an internal signal source. (Photo courtesy of Hewlett-Packard Company.)

TABLE 8-4 SPECIFICATIONS OF A TYPICAL NETWORK ANALYZER.

Frequency range: 5 Hz to 200 MHz
Measurement range: 0 to −130 dBm
Dynamic range: 100 dB
Amplitude accuracy: ±0.2 dB
Phase accuracy: ±2 degrees

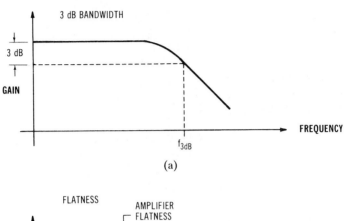

(a)

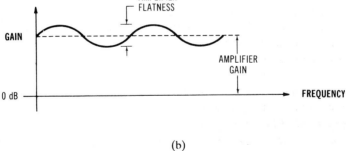

(b)

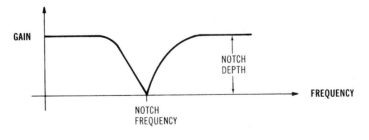

(c)

Figure 8-32 Examples of network analyzer measurements. The most common network analyzer display is gain in dB versus logarithmic frequency. (a) Measurement of 3-dB bandwidth. (b) Measurement of amplifier gain and flatness. (c) Measurement of filter notch depth and frequency.

The frequency response plot of a low-pass filter gives the complete response of the filter as well as allowing measurement of the 3-dB bandwidth (Figure 8-32a). The characteristics of an amplifier can also be measured (Figure 8-32b). In addition to the nominal gain of the circuit, the flatness of the response can easily be found. Rejection measurements, requiring a wide dynamic range, are common in electronic filter work (Figure 8-32c). The depth and frequency of a notch filter are measured.

8.12 DISTORTION ANALYZERS

Distortion analyzers measure the distortion produced in the circuit being tested. Basically, the distortion analyzer applies a very pure sine wave to the input of a circuit and measures the resulting output. If the circuit is distortion free, the output will also be a pure sine wave. However, in real circuits, some distortion may be introduced (typically harmonics). The distortion meter determines the amount of distortion in the output by filtering out the original sine wave and measuring the RMS voltage of all residual signals, including spurious responses and noise.

Figure 8-33 shows the simplified block diagram of a distortion analyzer. The sine wave output is connected to the circuit under test. (Not all distortion analyzers have internal sources.) This sine wave must be very pure since its distortion will show up as distortion at the output of the circuit and ultimately limit the accuracy of the instrument. The output of the circuit under test is connected to the input of the distortion analyzer. The analyzer uses a narrow notch filter to remove the original sine wave from the input signal. Anything left over is assumed to be distortion in the signal caused by the circuit being tested. The distortion is then compared to the waveform, including the original sine wave, and the distortion is read out on the meter as a percentage of the original signal. It is important to note that noise and spurious responses present in the circuit will also contribute to the distortion reading, since they will not be removed by the notch filter. Most distortion analyzers include selectable low-pass and high-pass filtering to help eliminate these unwanted signals.

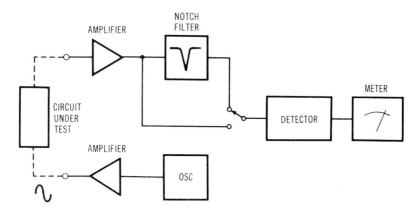

Figure 8-33 The conceptual block diagram of a distortion analyzer. The internal oscillator may be omitted.

Simple distortion analyzers work only at one frequency, typically 1 kHz. More advanced distortion analyzers provide the capability of selecting the test frequency. These instruments have a tunable notch filter which automatically tracks the frequency of the internal oscillator. The purity of the measured signal may be expressed in several different ways. For audio purposes, distortion (either in percent or dB relative to the fundamental) is usually preferred:

$$\text{distortion } (\%) = 100 \left(\frac{\text{noise} + \text{distortion}}{\text{signal}} \right)$$

Notice that noise present in the circuit will add to the measured distortion. Another alternative is expressing the imperfections in terms of either SINAD or SNR (signal-to-noise ratio).

$$\text{SINAD} = \frac{\text{signal} + \text{noise} + \text{distortion}}{\text{noise} + \text{distortion}}$$

$$\text{SNR} = \frac{\text{signal}}{\text{noise} + \text{distortion}}$$

SINAD measurements are often used to measure the sensitivity of a radio receiver. A signal generator with a 1-kHz modulating tone is connected to the antenna connection of the receiver. The signal generator's frequency is set to match the receiver frequency. A distortion meter is connected to the audio output of the receiver where it measures the SINAD of the recovered audio. The amplitude of the signal generator is reduced until a particular SINAD value is reached, typically 12dB. The signal generator amplitude at this point defines the sensitivity of the receiver.[3]

REFERENCES

ADAM, STEPHEN F. *Microwave Theory and Applications*. Englewood Cliffs, NJ: Prentice-Hall, Inc., 1969.

ENGELSON, MORRIS. *Modern Spectrum Analyzer Theory and Applications*. Dedham, MA: Artech House, Inc., 1984.

KINLEY, R. HAROLD *Standard Radio Communications Manual*, Englewood Cliffs, NJ: Prentice-Hall, Inc., 1985.

OLIVER, BERNARD M., and JOHN M. CAGE. *Electronic Measurements and Instrumentation*. New York: McGraw-Hill Book Company, 1971.

WITTE, ROBERT A. *Spectrum and Network Measurements*. Englewood Cliffs, NJ: Prentice-Hall, Inc., 1991.

[3]For more information on receiver testing see Kinley (1985).

chapter nine

Logic Probes
———————— *and Analyzers* ————————

Measurements of digital logic signals can be made with basic test instruments such as an oscilloscope or voltmeter. However, there are specialized instruments which can make logic measurements more convenient. The LOGIC PROBE is used to measure individual logic signals, while the LOGIC ANALYZER measures many digital signals simultaneously. Besides displaying the logic state of the signals, the logic analyzer functions as a data interpreter by displaying the captured logic signals in a variety of forms, including octal numbers, hexadecimal numbers, and microprocessor instruction codes.

9.1 LOGIC PROBES

A logic probe is a small instrument built into a probe-size case for checking out digital circuits (Figure 9-1). Generally, a logic probe indicates whether the voltage present corresponds to a logic HIGH or a logic LOW. The actual voltage of the signal is not displayed, since it is not significant except whether it's above or below the logic thresholds. Better probes also provide a pulse indicator, which flashes if logic pulses are present. Detailed information such as duty cycle or period cannot be accurately determined with a logic probe, but the presence or absence of a stream of bits can easily be found. The specifications of a typical logic probe are shown in Table 9-1.

9.1.1 Logic Thresholds

A logic probe must be designed to use the correct logic levels for measuring a particular circuit. These levels were defined in Chapter 1. The two most common digital logic technologies are TTL (transistor-transistor logic) and CMOS (comple-

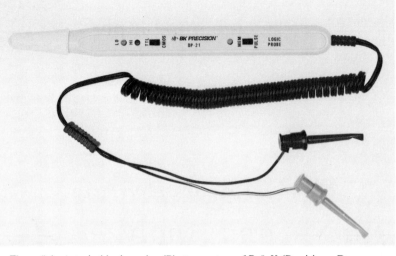

Figure 9-1 A typical logic probe. (Photo courtesy of B & K (Precision—Dynascan Corporation.)

mentary metal-oxide semiconductor). Sometimes these two are lumped together in the same category called "5 volt" logic. This is misleading since although both TTL and CMOS can operate using 5-volt supplies, the logic thresholds are *not* the same. To accommodate both technologies, many logic probes are switchable between TTL and CMOS. Regardless of the type of probe, it is important that the probe's logic thresholds match the logic family being tested.

9.1.2 Logic Probe Indicators

Different logic probes provide different types of logic state indication. The following list is roughly in order of most common to least common.

TABLE 9-1 SPECIFICATIONS FOR A TYPICAL LOGIC PROBE SHOW A FREQUENCY RANGE UP TO 50 MHZ AND DETECTION OF PULSES AS SHORT AS 10 NSEC.
Typical Logic Probe Specifications

Frequency range: 0 to 50 MHz
Minimum detectable pulse width: 10 nsec
Input impedance: 2 MΩ
Logic thresholds:
    TTL: High = 2.4 V, low = 0.8 V
    CMOS: High = 70% of supply voltage
               Low = 30% of supply voltage

1. *Logic high*: indicates that the voltage is greater than high-logic threshold; therefore, it is a valid HIGH-logic signal.

2. *Logic low*: indicates that the voltage is less than the low-logic threshold; therefore, it is a valid LOW-logic signal.

3. *High-impedance state*: indicates that the voltage is neither a logic low nor a logic high. Normally, this means that the digital gate is in the high-impedance state (tristate) or that the logic probe is not connected to the output of a gate (open circuit). The high-impedance state may be indicated by the absence of both the logic-high and the logic-low indicators.

4. *Pulse*: indicates that the voltage is changing from a valid low-logic level to a valid high-logic level (or vice versa). Often the regular logic-low and logic-high indicators are flashed on and off when a pulse occurs. The better logic probes provide pulse-stretching circuitry that catches very short pulses and lights the indicator long enough for the human eye to detect it. Without this special circuitry, short pulses might go undetected by the user, even though the logic level indicator was on for a brief time.

5. *Pulse memory (or pulse trap)*: indicates that a pulse has occurred since the last time the memory was cleared. This is very useful for cases where a digital pulse comes along infrequently but must still be detected. The probe is connected to the circuit and the clear button is pressed, clearing the pulse memory. The indicator stays off until a pulse comes along; then the indicator comes on and stays on until cleared again.

9.1.3 Oscilloscope versus Logic Probe

At first glance, it might seem that a logic probe would not be needed if an oscilloscope is available. This is somewhat true, but the logic probe does have some advantages over the oscilloscope. The oscilloscope gives more information than is required for troubleshooting a digital circuit. Having the voltage waveform displayed, including all of its imperfections, may make it more difficult to troubleshoot a logic circuit. With a logic probe, the user is sheltered from all the analog information but is given the digital information. The user does not have to memorize the logic thresholds of the particular technology being used, but instead just looks at the indicator on the probe to determine the logic state. Of course, if the analog behavior of the circuit is in question (such as too much overshoot or glitching), then an oscilloscope is quite useful.

The other advantage that the logic probe has over the oscilloscope is when very short pulses need to be detected. If a scope is used, the triggering and sweep controls must be adjusted to display the pulse properly. This may be tricky, particularly if the pulse occurs infrequently. If the scope does not have display storage capability, then when the pulse does finally occur it will be a brief flash across the screen.[1] With a logic probe, the logic levels are already set up so there is no triggering problem. The pulse indicator will detect pulses that are extremely short and display them long enough for

---

[1]Usually at just the time that the user blinks.

a human to see the indicator. If that is not sufficient, then a probe with pulse memory capability can be used to catch the pulse.

9.2 THE LOGIC ANALYZER

A logic probe provides a convenient means of viewing the state of a logic signal. Unfortunately, a logic probe can view only one signal at a time and, except for the pulse capture feature, cannot store a record of the logic signal over time. A LOGIC ANALYZER overcomes these objections by replicating the logic sensing feature of a logic probe, allowing many channels to be measured. A logic analyzer also stores the state of the logic inputs in memory so that a historical record can be displayed.

Since a conventional logic analyzer has two main operating modes, it can be thought of as two instruments in one: a timing analyzer and a state analyzer. Consider the simplified block diagram of a logic analyzer shown in Figure 9-2. The logic signals being measured enter the logic analyzer via probe buffers which present a high impedance to the circuits under test. The outputs of the probe buffers are clocked into a latch at regular intervals. The latched data are stored in memory and eventually displayed by the microprocessor.

Note that the signal which clocks the latch can come from two different sources. For state analysis, an external clock is used. This clock comes from the circuit under test and is often the master clock in the system. Each edge of the external clock causes the logic inputs to be stored into memory, producing a listing of logical information occurring at each clock state. For timing analysis, we want to monitor the logic signals on a more continuous basis, independent of the system clock. In timing analysis, a clock internal to the logic analyzer is used which clocks the latch continuously. This type of clocking scheme is referred to as ASYNCHRONOUS operation, since the internal analyzer clock runs asynchronous to the logic system being measured.

Since digital systems can run very fast with clock rates routinely above 10 MHz, it is easy to fill up the memory quickly. The CLOCK QUALIFIER input is used to force the logic analyzer to ignore certain clock cycles which are not needed. Similarly, the STORE CONTROL block can be programmed to store only data which meet certain requirements (such as a particular logic pattern) so that memory can be conserved.

The microprocessor reads the data from memory and writes it to the display. The data can be displayed in a variety of formats, as is discussed later in this chapter.

A typical logic analyzer is shown in Figure 9-3.

9.3 TIMING ANALYZER

When operated as a timing analyzer, the logic analyzer samples the waveform at each edge of its internal clock in a manner similar to a digitizing oscilloscope. One big difference is that the logic analyzer determines only whether the signal is a logic high or logic low, with no real voltage resolution. Figure 9-4 shows how a voltage waveform is sampled and recorded by a logic analyzer. For each sample interval, the logic analyzer displays the waveform as being high or low.

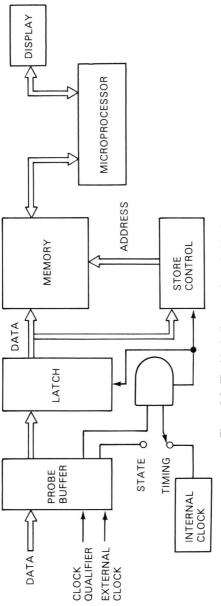

Figure 9-2 The block diagram of a typical logic analyzer.

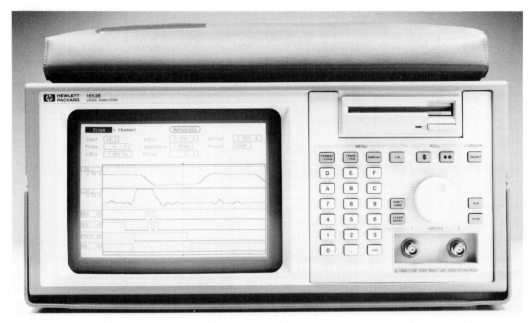

Figure 9-3 A typical logic analyzer with two oscilloscope channels. (Photo courtesy of Hewlett-Packard Company.)

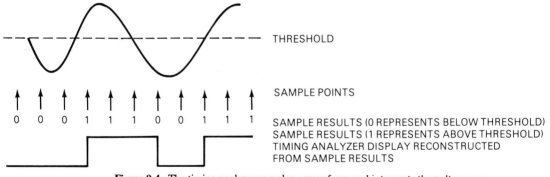

THRESHOLD

SAMPLE POINTS

0 0 0 1 1 1 0 0 1 1 1 1   SAMPLE RESULTS (0 REPRESENTS BELOW THRESHOLD)
SAMPLE RESULTS (1 REPRESENTS ABOVE THRESHOLD)
TIMING ANALYZER DISPLAY RECONSTRUCTED
FROM SAMPLE RESULTS

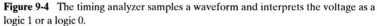

Figure 9-4 The timing analyzer samples a waveform and interprets the voltage as a logic 1 or a logic 0.

9.4 TIMING RESOLUTION

The logic analyzer samples the input signals only on the clock edges and cannot determine what happened in between its internal sample clock edges. This causes a timing uncertainty in the waveform as shown in Figure 9-5. For a clock frequency of f_{CLK}, the timing resolution (uncertainty) of measurement is $1/(f_{CLK})$. To maximize the timing resolution of a measurement, f_{CLK} needs to be as high as possible. The sample

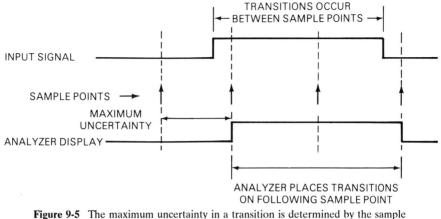

Figure 9-5 The maximum uncertainty in a transition is determined by the sample clock in the timing analyzer.

may be limited by the speed of the logic analyzer's internal circuitry. But there is a more important trade-off which takes place when choosing the clock frequency. The amount of time measured is affected by the clock frequency and the size of the memory.

$$T_{\text{MEAS}} = N \cdot T_{\text{CLK}}$$

where N is the size of the memory and T_{CLK} is the timing resolution ($1/f_{\text{CLK}}$). As f_{CLK} is increased, the timing resolution gets better, but the memory fills up faster.

Example 9-1.

What timing resolution is possible given that the total measurement time is 20 msec and the memory size is 1000?

$$T_{\text{MEAS}} = N \cdot T_{\text{CLK}}$$

so

$$T_{\text{CLK}} = \frac{T_{\text{MEAS}}}{N} = \frac{(0.020)}{1000} = 20 \; \mu\text{sec}$$

The available memory can be utilized more efficiently through the use of TRANSITIONAL SAMPLING. With transitional sampling, data are stored into memory only when the input signal changes logic state between samples. Storing the fact that a transition has occurred along with the time since the last transition allows the logic analyzer to reconstruct the timing waveform completely. For signals with frequently occurring transitions, the savings in memory is minimal, but for signals with multiple samples between transitions (such as data bursts), the amount of memory saved is large. We can think of this memory savings as a way to obtain better timing resolution for some given measurement time or, equivalently, a way to obtain longer measurement time while maintaining a certain timing resolution.

Figure 9-6 shows a measurement situation where transitional sampling saves a significant amount of memory without sacrificing timing resolution. The total measurement time is 50 μsec, and the desired timing resolution is 10 nsec. Normally, this

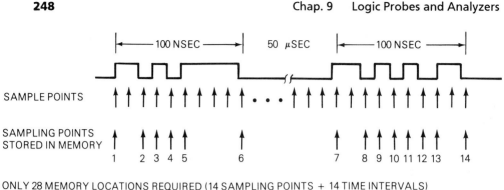

SAMPLE POINTS

SAMPLING POINTS
STORED IN MEMORY

1   2 3 4 5   6   7   8 9 10 11 12 13   14

ONLY 28 MEMORY LOCATIONS REQUIRED (14 SAMPLING POINTS + 14 TIME INTERVALS)

Figure 9-6 Transitional timing can produce significant savings in memory.

requires 50 μsec/10 nsec = 5000 memory locations. However, with transitional sampling, only 14 samples (and their timing information) are stored in memory, requiring a total of 28 memory locations.

9.5 GLITCH DETECT

Glitches in digital systems are a hardware designer's nightmare, and they occur all too often. It is usually difficult to identify glitches, let alone determine their cause. Logic analyzers often provide a special acquisition mode called GLITCH CAPTURE or GLITCH DETECT. In the case of a logic analyzer, a glitch is defined as any transition that crosses the logic threshold more than once between samples (Figure 9-7). Glitches are normally displayed with vertical lines which are easy to differentiate from the normal waveform. A special circuit is used to monitor the logic analyzer inputs to see if any double (or more) transitions occur between samples. There is some minimum glitch width which is detectable by the logic analyzer, typically a few nanoseconds.

The ability to capture and display a glitch helps identify glitches in the timing waveforms. Sometimes a glitch occurs very infrequently, so some logic analyzers provide GLITCH TRIGGER. This feature monitors the analyzer inputs and triggers the analyzer when a glitch occurs. This capability allows the user to set up the analyzer in a "watch dog" role, waiting for a glitch to occur. After the glitch triggers the logic

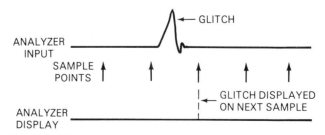

Figure 9-7 Glitches are defined as multiple transitions between samples.

analyzer, the captured waveforms can be analyzed to understand what caused the glitch.

9.6 DIGITAL TESTING EXAMPLE—4-BIT COUNTER

In Chapter 5, the operation of a simple 4-bit digital counter was used as an example of an oscilloscope measurement. A logic analyzer can also be used to check the operation of this circuit. Figure 9-8 (same as the Chapter 5 figure) shows the logic signals produced by the 4-bit counter. To display these timing waveforms on a logic analyzer, the four outputs (Q_0, Q_1, Q_2, and Q_3) and the clock are connected to the logic analyzer inputs. With the analyzer configured for timing analysis, the counter's clock is treated just like any other logic signal. The period and duty cycle of the clock can be checked within the limits of the timing resolution of the measurement. The logic analyzer display of these signals is shown in Figure 9-9.

Note that this measurement is quite similar to the oscilloscope measurement of Chapter 5. One difference is that most oscilloscopes would not have enough channels to display all five counter signals simultaneously. The number of channels becomes a

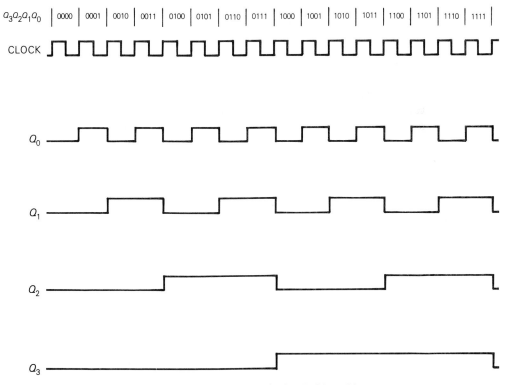

Figure 9-8 The waveforms associated with a 4-bit counter.

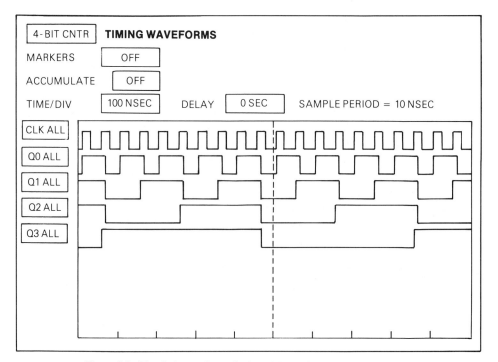

Figure 9-9   The timing analyzer display of the 4-bit counter's waveforms.

much bigger issue in practical digital systems where the address and data buses are often 16, 32, or 64 bits wide. Our simple 4-bit counter example is instructive, but requires very few channels compared to a typical digital system. Of course, one advantage the oscilloscope does have is that it can show the details of the waveform shape such as overshoot and rise time. Failures due to subtle imperfections in the waveform may not be uncovered by a logic analyzer since the logic analyzer is limited to measuring only logic high and logic low.

9.7 STATE ANALYZER

As previously mentioned, when operated as a state analyzer, the logic analyzer uses a signal from the circuit under test as the clock which determines when the logic signals are stored into memory. Often, the main clock of the digital system is used for the logic analyzer clock, but not always. Using a clock from the circuit under test means that the logic analyzer runs SYNCHRONOUS with the circuit and acquires the data at just the right time. It also means that the logic signals will not be measured in between these clock pulses regardless of how the logic analyzer is configured. (Compare this to timing analysis, where the logic analyzer's internal clock can be changed to view different amounts of time duration and to obtain a suitable timing resolution.)

A logic analyzer (configured for state analysis) can be used to check the operation of the 4-bit counter. The counter's clock signal can be used as the logic analyzer's clock as well since the counter changes state only on a clock edge. If the logic analyzer is set to clock on the rising edge (and the counter uses the rising edge), the logic analyzer will capture the logic state of the counter just when the clock goes high. This presents a quandary for some instrument users since the logic analyzer is trying to record the counter state just as the state is changing. A logic analyzer will record the clock state which was present just before the clock occurs. Since it takes a small finite time for the counter to change its state, the counter outputs will remain constant during the rising edge of the clock and for a very short time thereafter. Thus, the analyzer can reliably capture the logic state even though it is about to change.

The logic analyzer's display looks much different when doing state analysis. Instead of timing waveforms, the analyzer shows the user a list of logic states, each one acquired on an individual clock edge (Figure 9-10). Note that the state of the clock is not shown because the clock is always at the same state (low) just before a low-to-high transition. The simplest format to use in displaying the logic state is just binary (1 or 0). For measurement situations using many logic channels, binary patterns can be difficult to interpret, and other formats are used. (These are discussed later in the chapter.)

Measuring the 4-bit counter with both the timing analyzer and state analyzer shows the unique characteristics of both. If the user is interested in checking the timing

4-BIT CNTR	STATE LISTING
MARKERS	OFF

LABEL	>	Q
BASE	>	BIN
+ 0000		0000
+ 0001		0001
+ 0002		0010
+ 0003		0011
+ 0004		0100
+ 0005		0101
+ 0006		0110
+ 0007		0111
+ 0008		1000
+ 0009		1001
+ 0010		1010
+ 0011		1011
+ 0012		1100
+ 0013		1101
+ 0014		1110
+ 0015		1111

Figure 9-10 The state analyzer display of the 4-bit counter's waveform.

relationship of the digital signals, timing analysis is the clear choice. On the other hand, if the counter is used in a circuit such that its digital count has meaning, the state analyzer is preferred. The measurement system mirrors the use of the circuit.

9.8 MICROPROCESSOR MEASUREMENT

Choosing an appropriate clock signal for state analysis can be something of an art. The main system clock of a digital circuit is an obvious candidate, but it is not always the best choice. Figure 9-11 shows a mythical microprocessor system which includes a main clock, microprocessor, and memory. The timing of the memory read cycle is shown in Figure 9-12. The microprocessor first sets the read/write line (R/W) high to indicate that the next cycle is a read cycle (not a write cycle). Then the microprocessor outputs the address of the memory location to be read onto the address lines. After these signals have settled to their valid states, the active-low address strobe $(\overline{AS})^2$ goes low to indicate to the memory that the address is available. The memory responds by producing the data at that address on the data lines. A short time later, the microprocessor pulls the data strobe $(\overline{DS})$ low, then high, and reads the data into the microprocessor.

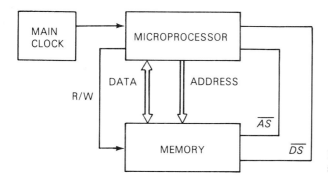

Figure 9-11 A mythical microprocessor system.

Now suppose we need to monitor the reads being done from the memory using a logic analyzer. If we need to check the timing of the various signals, the logic analyzer would be configured for timing analysis. But let's suppose that the timing of the circuit is just fine, but the data coming back from the memory are occasionally in error. In this case, state analysis is called for since we just need to monitor the address and data lines for each read cycle. So how do we connect the analyzer to the circuit? Clearly, we need to connect logic analyzer inputs to the address lines, the data lines, and perhaps the read/write line (so we can tell the difference between read and write cycles). One might be tempted to use the main clock as the clock input to the logic analyzer. Examining the rising (or even falling) edges of the main clock reveals that

---

[2]The overbar is used here to indicate that the logic signal is active low. That is, its main action occurs when the logic line goes into the low state.

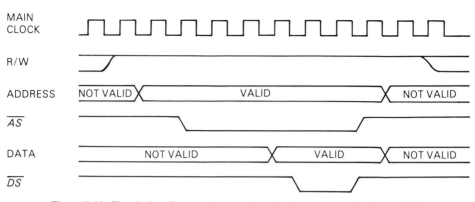

Figure 9-12 The timing diagram for a read cycle from memory in the mythical microprocessor system.

these edges occur at times when the address and data lines are not valid. So using the main clock as our analyzer clock would result in a state listing with many useless entries. Maybe the falling edge of address strobe would work since it occurs only once per memory cycle. But no, the data lines are not valid when the falling edge occurs. But we can clock on the rising edge of address strobe, since both the address and data are valid at that time.

If address strobe is used as the clock, write cycles would be mixed in with the read cycles. To prevent this, the read/write line can be used as a CLOCK QUALIFIER with the analyzer set to ignore clocks when the read/write line is in the write (low) state.

9.9 DATA FORMATS

As mentioned earlier, the simplest format for displayed digital signals is binary, with each logic analyzer input being a 1 or a 0. For measurements which have many channels, binary format can become unwieldy. Interpretation of high-channel-count measurements can be enhanced by displaying the data in octal or hexadecimal form.

Octal, as the name implies, uses base 8 arithmetic to group binary signals together in groups of three. A conversion table from binary to octal is given in Table 9-2.

Similarly, binary data can be presented in hexadecimal form, which is derived from base 16 arithmetic. With hexadecimal (also known as "hex"), four binary numbers are combined to produce one hexadecimal digit, as shown in Table 9-3.

To facilitate the exchange of alphanumeric information, the recommended USA Standard Code for Information Interchange (USASCII or more commonly ASCII) was developed. This code is used in a wide variety of computers, terminals, and related equipment and is the most common method for encoding textual information into digital form. Most logic analyzers allow the user to interpret and display ASCII encoded data. (The ASCII code is listed in Appendix E.) Besides the usual alphabetic

TABLE 9-2 OCTAL CONVERSION TABLE.

Binary	Octal
000	0
001	1
010	2
011	3
100	4
101	5
110	6
111	7

TABLE 9-3 HEXADECIMAL CONVERSION TABLE.

Binary	Hexadecimal
0000	0
0001	1
0010	2
0011	3
0100	4
0101	5
0110	6
0111	7
1000	8
1001	9
1010	A
1011	B
1100	C
1101	D
1110	E
1111	F

and numeric symbols, ASCII also includes data transmission symbols such as "Line feed" and "End of transmission."

9.10 STORAGE QUALIFICATION

Many logic analyzer measurements are an exercise in data reduction. With clock rates in the tens of MHz and 64 or more input channels, a logic analyzer is capable of acquiring large amounts of data in a short time. The challenge is sifting through the data to provide meaningful information. Often the logic analyzer's memory fills up before we have captured the desired event.

One feature that can help the user selectively acquire data is called STORAGE QUALIFICATION. Storage qualification allows the user to control the type of data which is stored into the logic analyzer memory. Earlier in the chapter, using the read/write line to qualify clocks and ignore write operations was discussed. We could just as easily use storage qualification to accomplish the same thing. The logic analyzer could be configured to store only logic patterns which have the read/write line HIGH.

Suppose that we discover that the problem with reading the memory in our mythical microprocessor system occurs only within a certain block of addresses. The logic analyzer can be set to store only acquisitions which have the upper address bits set for that particular block of memory. We can also require that the read/write line be HIGH, limiting the stored data to read operations. Alternatively, the read errors may be occurring at all addresses, but only when a certain data pattern is read. If so, the analyzer can be configured to store data on all addresses, but only when the data fits the troublesome data pattern. Using storage qualification lets the troubleshooter narrow the problem down until the faulty logic operation is isolated.

9.11 TRIGGERING AND SEQUENCE LEVELS

If we just tell the logic analyzer to run without any special triggering, it begins storing data on the next clock. Another aid in solving the data reduction problem is to specify a pattern to trigger on. Suppose our memory read problem occurs only after memory location 000000 has been read. We could set the logic analyzer to trigger on a read cycle with the address set to 000000. This means that the analyzer will not trigger or store any data until it sees a read cycle from that location; then it will begin storing data according to whatever storage qualification is requested (if any). Together, pattern triggering and storage qualification can be used to isolate a problem in a digital circuit.

There is one more level of control available in most logic analyzers called SEQUENCING. Sequencing expands the concept of triggering on a pattern to include finding patterns in a particular sequence. Suppose that our memory read problem is particularly subtle since errors occur in the data only when address 000001 is read, but only after location 000000 is read. To trace on this sequence, we would set the analyzer to first find a read from address 000000 and then trigger on a read from 000001.

Sequencing and triggering capability depends on the particular model of logic analyzer, but the general techniques already discussed are included in a modern logic analyzer. The exact method of specifying these capabilities also varies but is explained in the analyzer's user manual.

9.12 MICROPROCESSOR PROGRAM FLOW

One of the most common applications of logic analyzers is to monitor the program flow in a microprocessor system. The address, data, status, and control lines of the microprocessor are monitored to keep track of what the microprocessor is doing. The processor can be fetching instructions from memory, reading data from memory, writing data to memory, or handling interrupts. Sometimes the logic analyzer is used to debug a problem with the microprocessor software, and sometimes it is used to uncover a hardware problem. Sometimes the problem is unknown in origin (hardware or software) or may be a subtle hardware-software interaction. In all these cases, a logic analyzer can help track down and isolate the problem.

When debugging a microprocessor system, the sequencing and triggering capabilities are used to track the software program flow. For example, suppose we want to trace the execution of a low-level math subroutine, but only when it is called from a particular section of high-level code. We set the logic analyzer first to find an instruction being executed at an address known to be in the high-level code and then trigger on the entry point of the math routine. The logic analyzer stores only the processor activity starting with the entry to the math routine, saving the analyzer memory for when it's really needed.

The standard probes which come with the logic analyzer can be used to connect to the microprocessor signals, but special adapters called PREPROCESSORS are more convenient. Preprocessors typically plug into the microprocessor socket, and the microprocessor plugs into the preprocessor. A preprocessor provides a reliable, fast, and convenient way of connecting to the circuit under test. In addition, it provides clocking and demultiplexing circuits to capture the signals to and from the processor more reliably. Additional status lines may be decoded in the preprocessor, giving the user more information about the internal state of the microprocessor. (Preprocessors are also available which handle standard buses and interfaces such as IEEE-488, RS-232C, RS-449, SCSI, VME, and VXI.)

Even with a preprocessor, the address, data, status, and control signals associated with a microprocessor can be difficult to interpret. The logic analyzer user must have a good understanding of the microprocessor's instruction set to decode the bit patterns that the logic analyzer displays. Even with this understanding, the logic analyzer user must look up which instruction corresponds to a particular logic pattern. Logic analyzer manufacturers offer INVERSE ASSEMBLERS (also known as DISASSEMBLERS) which perform this task for the user. The inverse assembler consists of software which runs in the logic analyzer and interprets the instructions and data captured by the analyzer. The program flow is displayed in the microprocessor's assembly language format. The combination of an inverse assembler and preprocessor provides a powerful analysis capability for microprocessor system development. Note that inverse assemblers and preprocessors are designed to work for a particular model of microprocessor. Figure 9-13 shows a typical inverse assembler listing. Shown left to right across the listing are the address of the memory access, the instruction being executed, the operand associated with the instruction, the hex pattern on the data bus, and an indication of whether a read or write is being performed to memory.

9.13 LOGIC ANALYZER PROBING

Logic analyzer inputs are connected to the circuit under test using the probes provided with the analyzer. These probes are usually arranged in a pod containing 8 to 16 input channels. Like all measurement instruments, logic analyzers disturb the circuit under

M68332EVS	**STATE LISTING**	(INVASM)

MARKERS   OFF

LABEL >	ADDR	68010/332 MNEMONIC		DATA _	R/W
BASE >	HEX	HEX		HEX	SYMBOL
+ 0058	6DA4C	MOVE .L	D0, − [A7]	2F00	RD
+ 0059	6DA4E	MOVEQ.L	#00000001, D0	7001	RD
+ 0060	6DA50	ST.B	D0	50C0	RD
+ 0061	02FF4	0000	DATA WRITE	0000	WR
+ 0062	02FF6	0180	DATA WRITE	0180	WR
+ 0063	6DA52	LEA.L	000000, A0	41F8	RD
+ 0064	6DA54	0000	PGM READ	0000	RD
+ 0065	6DA56	LEA.L	4000 [A0], A1	43E8	RD
+ 0066	6DA58	4000	PGM READ	4000	RD
+ 0067	6DA5A	LEA.L	0400 [A0], A5	4BE8	RD
+ 0068	6DA5C	0400	PGM READ	0400	RD
+ 0069	6DA5E	MOVEQ.L	#00000000, D0	7000	RD
+ 0070	6DA60	MOVE.L	[A7] +, D0	201F	RD
+ 0071	6DA62	RTS		4E75	RD
+ 0072	02FF4	0000	DATA READ	0000	RD
+ 0073	02FF6	0180	DATA READ	0180	RD

Figure 9-13   An inverse assembler listing for the Motorola 68332 microprocessor.

test when the analyzer is attached, so the amount of loading the probe presents to the circuit is important. The concepts of resistive and capacitive loading, discussed in Chapter 4 with respect to oscilloscopes, also apply to logic analyzer probes. A typical logic analyzer probe presents a load of 100 kΩ and 8 pF to the circuit under test. A load of 100 kΩ is large enough to avoid significant resistive loading effects on common digital circuits. The 8 pF of capacitive loading will introduce a small timing error, typically 1.3 nsec on a high-speed CMOS gate.

Connecting up 50 or 60 channels of logic analyzer probes is a time-consuming task. Logic analyzer manufacturers provide a variety of clips and adapters which help to varying degrees. Designers of digital circuits often include a socket on their circuit board which connects the important signals on the board directly to a logic analyzer. Such foresight can pay off in the long run via decreased troubleshooting time.

9.14 COMBINED SCOPE AND LOGIC ANALYZER

Sometimes it is a difficult choice whether to use a scope or a logic analyzer to attack a difficult digital problem. The scope generally provides better timing resolution and a better measure of waveform fidelity while the logic analyzer wins on number of channels and manipulation of displayed data. Instrument manufacturers have solved this dilemma by developing logic analyzers which have built-in oscilloscopes. These built-in oscilloscopes are digital with sufficiently high sample rate so that fast single-shot events can be captured.

The scope-analyzer combination produces time-correlated results which can be displayed separately or simultaneously. Cross-triggering is also available, which allows the pattern trigger and sequencing capability of the logic analyzer to be used to arm or trigger the scope. The logic analyzer can be triggered by the scope as well. Complex interactions among digital hardware, software, and even analog hardware can be captured with such an instrument.

REFERENCES

"Accessories for HP Logic Analyzers." Hewlett-Packard Company, Publication Number 5952-3287, June 1990, Palo Alto, CA.

"Basic Concepts of Logic Analysis." Tektronix, Inc., February 1989, Beaverton, OR.

"Feeling Comfortable with Logic Analyzers." Hewlett-Packard Company, Publication Number 5954-2686, April 1988, Palo Alto, CA.

"Powerful, Affordable Logic Analysis." Hewlett-Packard Company, Publication Number 5959-3127, October 1989, Palo Alto, CA.

Table of Electrical Parameters, Units, and Standard Abbreviations

Quantity	Symbol	Unit	Abbreviation	Comments
charge	Q	coulomb	C	
current	I	ampere	A	ampere = coulomb per second
voltage	V (or E)	volt	V	
resistance	R	ohm	Ω	ohm = volt per ampere
power	P	watt	W	watt = volt-ampere
capacitance	C	farad	F	
inductance	L	henry	H	
frequency	f	hertz	Hz	hertz = cycle per second

Fundamental units (shown) can be modified by using the following standard prefixes:

Prefix	Multiplier	Symbol
femto	10^{-15}	f
pico	10^{-12}	p
nano	10^{-9}	n
micro	10^{-6}	μ
milli	10^{-3}	m
kilo	10^{3}	k
mega	10^{6}	M
giga	10^{9}	G

Examples:

$$2.5 \text{ MHz} = 2.5 \times 10^6 = 2,500,000 \text{ Hz}$$
$$150 \text{ } \mu\text{F} = 150 \times 10^{-6} = 0.000150 \text{ F}$$
$$25 \text{ mV} = 25 \times 10^{-3} = 0.025 \text{ V}$$

Table of Percentage Error and Decibel Relationships (for Voltage and Current)

Error (± percent)	Error(dB)		Error (dB, relative to ideal value)
0.01%	+0.00087	−0.00087	−80.00
0.02	+0.00174	−0.00174	−73.98
0.03	+0.00261	−0.00261	−70.46
0.04	+0.00347	−0.00348	−67.96
0.05	+0.00434	−0.00434	−66.02
0.06	+0.00521	−0.00521	−64.44
0.07	+0.00608	−0.00608	−63.10
0.08	+0.00695	−0.00695	−61.94
0.09	+0.00781	−0.00782	−60.92
0.10	+0.00868	−0.00869	−60.00
0.20	+0.0174	−0.0174	−53.98
0.30	+0.0260	−0.0261	−50.46
0.40	+0.0347	−0.0348	−47.96
0.50	+0.0433	−0.0435	−46.02
0.60	+0.0520	−0.0523	−44.44
0.70	+0.0606	−0.0610	−43.10
0.80	+0.0692	−0.0698	−41.94
0.90	+0.0778	−0.0785	−40.92
1.00	+0.0864	−0.0873	−40.00
2.00	+0.172	−0.175	−33.98
3.00	+0.257	−0.265	−30.46
4.00	+0.341	−0.355	−27.96
5.00	+0.424	−0.446	−26.02
6.00	+0.506	−0.537	−24.44
7.00	+0.588	−0.630	−23.10
8.00	+0.668	−0.724	−21.94
9.00	+0.749	−0.819	−20.92

10.00	+0.828	−0.915	−20.00
15.00	+1.214	−1.412	−16.48
20.00	+1.584	−1.938	−13.98
25.00	+1.938	−2.499	−12.04
30.00	+2.279	−3.098	−10.46
35.00	+2.607	−3.742	−9.12
40.00	+2.923	−4.437	−7.96
45.00	+3.227	−5.193	−6.94
50.00	+3.522	−6.021	−6.02

This table is valid for both voltage and current values. For power calculations, the dB values must be multiplied by 1/2.

If a particular instrument reading is thought of as consisting of two parts, the ideal value and the error in the reading, then

$$reading = ideal\ value + error$$

The error (%) value is the error expressed as a percentage of the ideal value

$$error\ (\%) = 100 \left(\frac{error}{ideal\ value} \right)$$

The error (dB) value is the amount by which the reading differs from the ideal value, expressed in dB.

$$error\ (dB) = 20 \log \left(1 + \frac{error\ (\%)}{100} \right)$$

The error (dB, relative to ideal value) expresses the error in dB, relative to the ideal value.

$$error\ (dB,\ rel\ to\ ideal) = 20 \log \left(\frac{error\ (\%)}{100} \right)$$

Example:

A reading with a 1 percent error will be off by +0.0864 dB or −0.0873 dB, depending on the direction of the error. The error is −40 dB relative to the ideal value.

appendix c

List of Instrument
Manufacturers

B & K Precision—Dynascan
Corporation
6460 West Cortland Street
Chicago, IL 60635

Beckman Instruments, Inc.
2500 Harbor Boulevard
Fullerton, CA 92634

Hewlett-Packard Co.
1820 Embarcadero Road
Palo Alto, CA 94303

Hickok Electrical Instrument Co.
10514 Dupont Avenue
Cleveland, OH 44108

Iwatsu Instruments, Inc.
430 Commerce Boulevard
Carlstadt, NJ 07072

John Fluke Manufacturing Co.
P.O. Box C9090
Everett, WA 98206

Keithley Instruments, Inc.
28775 Aurora Road
Cleveland, OH 44139

Leader Instrument Corporation
380 Oser Avenue
Hauppauge, NY 11788

Marconi Instruments
3 Pearl Court
Allendale, NJ 07401

Panasonic Industrial Company
One Panasonic Way
Secaucus, NJ 07094

Simpson Electric Company
853 Dundee Avenue
Elgin, IL 60120

Tektronix, Inc.
P.O. Box 1700
Beaverton, OR 97075

Triplett Corporation
One Triplett Drive
Bluffton, OH 45817

Wavetek Corporation
9191 Towne Centre Drive,
Suite 450
San Diego, CA 92122

Mathematical Derivations
—————— *of Equations* ——————

AVERAGE (MEAN) VALUE OF A WAVEFORM

The average or mean value of a periodic waveform is found by integrating the waveform over one period.

$$\text{average of } x(t) = \frac{1}{T} \int_{t_0}^{t_0 + T} x(t)\, dt$$

where

T = the period
t_0 = the start of the integration which is arbitrarily chosen

RMS VALUE OF WAVEFORM

The RMS (root-mean-square) value of a waveform is calculated by squaring the waveform, finding its average value, and taking the square root.

$$V_{\text{RMS}} = \sqrt{\text{average of } v^2(t)}$$

$$= \sqrt{\frac{1}{T} \int_{t_0}^{t_0 + T} v^2(t)\, dt}$$

Example

$$v(t) = V_{0-p} \sin (2\pi ft)$$

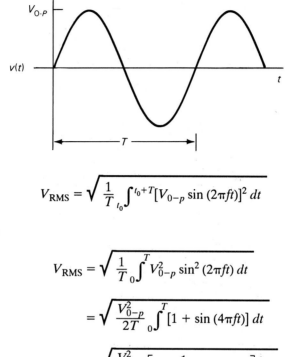

$$V_{\text{RMS}} = \sqrt{\frac{1}{T} \int_{t_0}^{t_0+T} [V_{0-p} \sin (2\pi ft)]^2 \, dt}$$

Choose $t_0 = 0$.

$$V_{\text{RMS}} = \sqrt{\frac{1}{T} \int_0^T V_{0-p}^2 \sin^2 (2\pi ft) \, dt}$$

$$= \sqrt{\frac{V_{0-p}^2}{2T} \int_0^T [1 + \sin (4\pi ft)] \, dt}$$

$$= \sqrt{\frac{V_{0-p}^2}{2T} \left[t - \frac{1}{4\pi f} \cos (4\pi ft) \right] \Big|_0^t}$$

$$= \sqrt{\frac{V_{0-p}^2}{2T} [T]}$$

$$V_{\text{RMS}} = \frac{V_{0-p}}{\sqrt{2}} \qquad \text{(sine wave)}$$

FULL-WAVE RECTIFIED AVERAGE VALUE OF A WAVEFORM

The full-wave rectified average value of a periodic waveform is found by taking the average of the absolute value of the waveform.

$$V_{\text{AVG}} = \text{average of } |v(t)|$$

$$= \frac{1}{T} \int_{t_0}^{t_0+T} |v(t)| \, dt$$

Example:

$$v(t) = V_{0\text{-}p} \sin (2\pi ft)$$

Choose $t_0 = 0$.

$$V_{\text{AVG}} = \frac{1}{T} \int_0^T | V_{0\text{-}p} \sin (2\pi ft) | \, dt$$

Since $|V_{0\text{-}P} \sin 2\pi ft|$ is the same for the time periods of 0 to $\dfrac{T}{2}$ and $\dfrac{T}{2}$ to T.

$$V_{\text{AVG}} = \frac{1}{T} 2 \int_0^{\frac{T}{2}} | V_{0\text{-}p} \sin (2\pi ft) | \, dt$$

$$= \frac{2V_{0\text{-}p}}{T} \int_0^{\frac{T}{2}} \sin (2\pi ft) \, dt$$

$$= \frac{2V_{0\text{-}p}}{T} \left[- \frac{1}{2\pi f} \cos (2\pi ft) \right] \Big|_0^{\frac{T}{2}}$$

$$= \frac{2V_{0\text{-}p}}{T} \left(\frac{1}{2\pi f} + \frac{1}{2\pi f} \right)$$

$$= \frac{2V_{0\text{-}p}}{\pi fT}$$

since $f = \dfrac{1}{T}$,

$$V_{\text{AVG}} = \frac{2}{\pi} V_{0\text{-}P} \quad \text{(sine wave)}$$

BANDWIDTH AND RISE TIME FOR A SINGLE-POLE SYSTEM

A system with a single pole in the frequency domain has a frequency response of the form

$$H(f) = \frac{H_0}{1 + j\left(\dfrac{f}{BW} \right)}$$

where

    BW = the 3-dB bandwidth
    H_0 = the DC gain

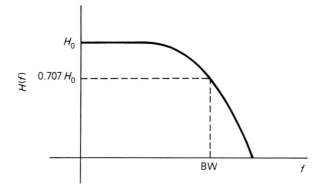

In the time domain, such a system has an exponential step response (assuming $H_0 = 1$)

$$v(t) = 1 - e^{t/\tau} \qquad t \geq 0$$

where τ is the time constant and

$$\tau = \frac{1}{2\pi \mathrm{BW}}$$

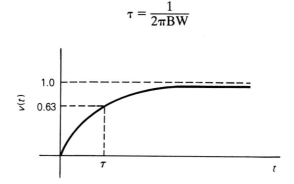

The rise time is defined as the time between the 10 percent point, t_1, and the 90 percent point, t_2.

$$0.1 = 1 - e^{-t_1/\tau}$$

$$t_1 = -\tau \ln (0.9)$$

$$t_2 = -\tau \ln (0.1)$$

The rise time, t_r, is

$$t_r = t_2 - t_1 = -\tau (\ln 0.1 - \ln 0.9)$$

$$= 2.197\tau$$

$$= 2.197 \left(\frac{1}{2\pi \mathrm{BW}} \right)$$

$$t_r = \frac{0.35}{\mathrm{BW}}$$

This relationship is exact for a single-pole system and is a good approximation for systems that have very little overshoot in their step responses.

FREQUENCY RESPONSE DUE TO COUPLING CAPACITOR

The effect of a coupling capacitor can be modeled as

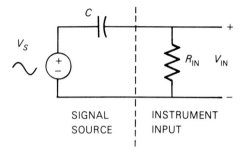

$$V_{IN} = V_S \left(\frac{R_{IN}}{R_{IN} + \dfrac{1}{j2\pi f C}} \right)$$

$$\frac{V_{IN}}{V_S} = \frac{j2\pi f c R_{IN}}{j2\pi f c R_{IN} + 1}$$

At $f = 0$ (DC),

$$\frac{V_{IN}}{V_S} = 0$$

For large f (high frequency),

$$\frac{V_{IN}}{V_S} = 1$$

At $f = \dfrac{1}{2\pi R_{IN} C}$

$$\left| \frac{V_{IN}}{V_S} \right| = \frac{1}{\sqrt{2}} = 0.707$$

So the 3-dB frequency of this system is

$$f_{3dB} = \frac{1}{2\pi R_{IN} C}$$

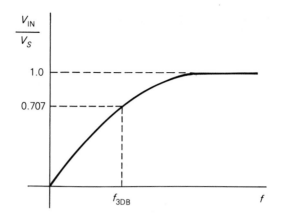

POLAR AND RECTANGULAR FORMATS

A vector number can be represented in both polar and rectangular form.

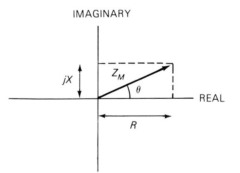

rectangular format : $z = R + jx$
polar format: $z = Z_m \angle \theta$

where

$$Z_m = \sqrt{R^2 + X^2}$$

$$\theta = \tan^{-1}\left(\frac{X}{R}\right) \quad \text{for } R \geq 0$$

$$= \tan^{-1}\left(\frac{X}{R}\right) + 180 \text{ deg} \quad \text{for } R < 0, X \geq 0$$

$$= \tan^{-1}\left(\frac{X}{R}\right) - 180 \text{ deg} \quad \text{for } R < 0, X < 0$$

Other relationships used for converting polar to rectangular format are

$$R = Z_M \cos \theta$$

$$X = Z_M \sin \theta$$

RESONANT FREQUENCY

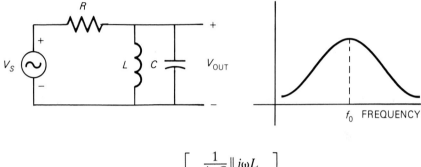

$$V_{OUT} = V_S \left[\frac{\dfrac{1}{j\omega C} \| j\omega L}{R + \dfrac{1}{j\omega C} \| j\omega L} \right]$$

$$= V_S \left[\frac{\dfrac{(j\omega L)}{(1 - \omega^2 LC)}}{R + \left(\dfrac{j\omega L}{1 - \omega^2 LC} \right)} \right]$$

$$= V_S \left[\frac{j\omega L}{R(-\omega^2 LC) + j\omega L} \right]$$

V_{OUT} reaches a maximum when the real part of the denominator is zero:

$$R(1 - \omega_0^2 LC) = 0$$

$$\omega_0^2 LC = 1$$

Resonance occurs when

$$\omega_0 = \frac{1}{\sqrt{LC}}$$

$$f_0 = \frac{\omega}{2\pi} = \frac{1}{2\pi\sqrt{LC}}$$

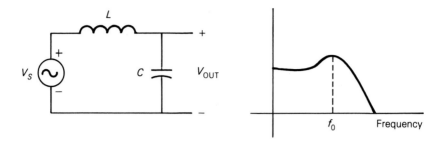

$$V_{\text{OUT}} = V_S \left[\frac{\dfrac{1}{j\omega C}}{j\omega L + \dfrac{1}{j\omega C}} \right]$$

$$= V_S \left(\frac{1}{1 - \omega^2 LC} \right)$$

V_{OUT} is maximum when the denominiator is zero:

$$1 - \omega_0^2 LC = 0$$

Resonance occurs when

$$\omega_0 = \frac{1}{\sqrt{LC}}$$

$$f_0 = \frac{\omega_0}{2\pi} = \frac{1}{2\pi\sqrt{LC}}$$

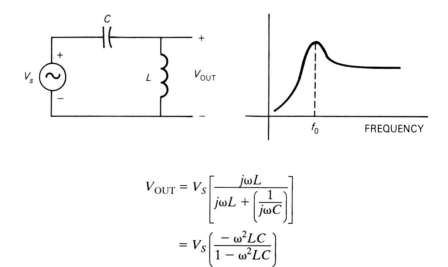

$$V_{\text{OUT}} = V_S \left[\frac{j\omega L}{j\omega L + \left(\dfrac{1}{j\omega C} \right)} \right]$$

$$= V_S \left(\frac{-\omega^2 LC}{1 - \omega^2 LC} \right)$$

V_{OUT} is maximum when the denominator is zero:

$$1 - \omega_0^2 LC = 0$$

Resonsance occurs when

$$\omega_0 = \frac{1}{\sqrt{LC}}$$

$$f_0 = \frac{\omega_0}{2\pi} = \frac{1}{2\pi\sqrt{LC}}$$

appendix e

ASCII Conversion Table

Binary	Hexadecimal	Decimal	ASCII	Comments
0000000	00	0	NUL	Null(All Zeros)
0000001	01	1	SOH	Start of Heading
0000010	02	2	STX	Start of Text
0000011	03	3	ETX	End of Text
0000100	04	4	EOT	End of Transmission
0000101	05	5	ENQ	Enquiry
0000110	06	6	ACK	Acknowledge
0000111	07	7	BEL	Bell or Alarm
0001000	08	8	BS	Backspace
0001001	09	9	HT	Horizontal Tabulation
0001010	0A	10	LF	Line Feed
0001011	0B	11	VT	Vertical Tabulation
0001100	0C	12	FF	Form Feed
0001101	0D	13	CR	Carriage Return
0001110	0E	14	SO	Shift Out
0001111	0F	15	SI	Shift In
0010000	10	16	DLE	Data Link Escape
0010001	11	17	DC1	Device Control 1
0010010	12	18	DC2	Device Control 2
0010011	13	19	DC3	Device Control 3
0010100	14	20	DC4	Device Control 4
0010101	15	21	NAK	Negative Acknowledge
0010110	16	22	SYN	Synchronous Idle
0010111	17	23	ETB	End of Transmission Block
0011000	18	24	CAN	Cancel
0011001	19	25	EM	End of Medium
0011010	1A	26	SUB	Substitute
0011011	1B	27	ESC	Escape
0011100	1C	28	FS	File Separator

Binary	Hexadecimal	Decimal	ASCII	Comments
0011101	1D	29	GS	Group Separator
0011110	1E	30	RS	Record Separator
0011111	1F	31	US	Unit Separator
0100000	20	32	SP	Space
0100001	21	33	!	
0100010	22	34	"	
0100011	23	35	#	
0100100	24	36	$	
0100101	25	37	%	
0100110	26	38	&	
0100111	27	39	'	
0101000	28	40	(	
0101001	29	41	)	
0101010	2A	42	*	
0101011	2B	43	+	
0101100	2C	44	'	
0101101	2D	45	−	
0101110	2E	46	.	
0101111	2F	47	/	
0110000	30	48	0	
0110001	31	49	1	
0110010	32	50	2	
0110011	33	51	3	
0110100	34	52	4	
0110101	35	53	5	
0110110	36	54	6	
0110111	37	55	7	
0111000	38	56	8	
0111001	39	57	9	
0111010	3A	58	:	
0111011	3B	59	;	
0111100	3C	60	<	
0111101	3D	61	=	
0111110	3E	62	>	
0111111	3F	63	?	
1000000	40	64	@	
1000001	41	65	A	
1000010	42	66	B	
1000011	43	67	C	
1000100	44	68	D	
1000101	45	69	E	
1000110	46	70	F	
1000111	47	71	G	
1001000	48	72	H	
1001001	49	73	I	
1001010	4A	74	J	
1001011	4B	75	K	
1001100	4C	76	L	
1001101	4D	77	M	
1001110	4E	78	N	
1001111	4F	79	O	
1010000	50	80	P	
1010001	51	81	Q	

Binary	Hexadecimal	Decimal	ASCII	Comments
1010010	52	82	R	
1010011	53	83	S	
1010100	54	84	T	
1010101	55	85	U	
1010110	56	86	V	
1010111	57	87	W	
1011000	58	88	X	
1011001	59	89	Y	
1011010	5A	90	Z	
1011011	5B	91	[	
1011100	5C	92	\	
1011101	5D	93	]	
1011110	5E	94	^	
1011111	5F	95	_	
1100000	60	96	`	
1100001	61	97	a	
1100010	62	98	b	
1100011	63	99	c	
1100100	64	100	d	
1100101	65	101	e	
1100110	66	102	f	
1100111	67	103	g	
1101000	68	104	h	
1101001	69	105	i	
1101010	6A	106	j	
1101011	6B	107	k	
1101100	6C	108	l	
1101101	6D	109	m	
1101110	6E	110	n	
1101111	6F	111	o	
1110000	70	112	p	
1110001	71	113	q	
1110010	72	114	r	
1110011	73	115	s	
1110100	74	116	t	
1110101	75	117	u	
1110110	76	118	v	
1110111	77	119	w	
1111000	78	120	x	
1111001	79	121	y	
1111010	7A	122	z	
1111011	7B	123	{	
1111100	7C	124			
1111101	7D	125	}	
1111110	7E	126	~	
1111111	7F	127	DEL	Delete

Index